Vortex Publishing LLC.
4101 Tates Creek Centre Dr
Suite 150- PMB 286
Lexington, KY 40517

www.vortextheory.com

© Copyright 2019 Vortex Publishing

All rights reserved. No part of this book may be reproduced or transmitted in any form or by any means, electronic or mechanical, including photocopying, recording or by any information storage and retrievable system without the prior written permission by the Publisher. For permission requests, contact the publisher.

Printed in the United States of America

1 2 3 4 5 6 7 8 9 10

Library of Congress Control Number: 2019953089

ISBN   978-1-7332996-2-6
eISBN 978-1-7332996-3-3

Editor's note: *All drawings in this book are original illustrations made by Dr. Moon. They are kept as they are to maintain the integrity of his work.*

TABLE OF CONTENTS

Introduction III
Prof. Dr. Victor V. Vasiliev's forward to *The End of the Concept of 'Time'!* IV

PART I
THE TRUE VISION OF THE UNIVERSE

Chapter 1: The Mystery Amid the Foundations of Science 1
Chapter 2: The Problem Is Time! 5
Chapter 3: The Undiscovered Territory 9
Chapter 4: The True Vision of Space 10
Chapter 5: The Proton and the Electron 15
Chapter 6: The Vortex 19
Chapter 7: The Neutron 24
Chapter 8: Particle Collisions 27
Chapter 9: The True Vision of Energy 31
Chapter 10: Creating the Forces of Nature 37
Chapter 11: The Force of Gravity 38
Chapter 12: The Electromagnetic Force 42
Chapter 13: The Weak Force 45
Chapter 14: The Strong Force 47
Chapter 15: The Anti-gravity Force Using Buoyancy as an Example 49
Chapter 16: Mass…[There Is No Higgs Boson Particle] 51
Chapter 17: The Terminal Velocity of Atoms 56
Chapter 18: The Shocking Truth About Time! 57
Chapter 19: How the Phenomenon of Time Is Created 58
Chapter 20: Time Dilation Is Finally Explained 60
Chapter 21: The True Vision of the Universe 62
Chapter 22: At Last – The Magic Cipher 63

PART II
GREAT MYSTERIES OF THE UNIVERSE FINALLY EXPLAINED

#1 The explanation of the particle and wave theory of light 64
#2 The explanation of the particle and wave theory of matter 64
#3 The explanation of the double slit interference patterns created by light and matter 66
#4 The explanation of ½ spin of particles 69
#5 Explanation of Inertia 72
#6 The explanation of Newton's three laws of motion 74
#7 Momentum 77
#8 The explanation of the Conservation of Momentum 77
#9 The explanation of the Conservation of Angular Momentum 78
#10 The explanation of the Conservation of Charge 80
#11 The explanation of the conservation of mass and energy 81
#12 The explanation of the great Mystery of Entropy 81
#13 The explanation of Dark "Matter" 82
#14 Dark Energy 85
#15 Anti-gravity Engineering 86
#16 What the neutrino really is and why it possesses both matter and energy characteristics 87

#17 The explanation of Buoyancy 88
#18 The explanation of the Ionic and Covalent Bonds in Chemistry 89
#19 The Mathematical Explanation of the Michelson Morley Experiment 91
#20 The Explanation of: The Strong force, The Weak force, The Electromagnetic Force, The force of Gravity and the Anti-gravity Force 92
#21 The Explanation of Mass 92
#22 The Explanation of Energy 92
#23 What causes acceleration 93
#24 The Explanation of the Muon's Prolonged Lifetime when moving at relativistic velocities 93
#25 All of the phenomenon associated with the Theory of Relativity are now explained 94
#26 Why all Charged "Particles" possess the same amount of Charge 94
#27 The explanation of Black Holes 95
#28 The explanation of Planck's Constant 96
#29 The explanation of high velocities and increasing mass 96
#30 The explanation of a striking parallel between Newton's Law of gravity and Coulomb's Law 97
#31 The explanation of the creation of the universe 98
#32 The explanation of time and time dilation effects 99
#33 The Grand Unification Theory? I don't think so! 99
#34 A neat trick of nature! Why matter does not flow in the rivers of flowing space! 100
#35 The reason why electrons "orbit" protons 101
#36 GOD! Have we discovered GOD! 102
#37 Universal Religion 103

PART III
ADDENDUM

Just prior to publication: A Blockbuster Discovery of enormous importance was made:
The Explanation of the Constant of Fine Structure 105
Chapter 2A: Creation of the 21 centimeter line 114
Chapter 3A: The explanation of the twin fine lines of hydrogen 123
Chapter 4A: "Electrostatics" is a mistake: it should be "Electrodynamics" !!! 125
Chapter 5A: "Electrodynamics" explains the ejection of alpha particles, gamma rays, and x-rays 127

References

National/International Conferences attended and
peer reviewed scientific papers presented 133

Books by author {A} and work presented in other published books/booklets 137

Other References 138

Subjects found in Book 3 .... **140**

INTRODUCTION

The Vortex Theory of Atomic Particles is the second volume in this six-part series. The first book, *The End of the Concept of "Time!"* tells how and why the Vortex Theory was discovered. Because this book was written for the general public, only the revolutionary scientific discoveries necessary to explain the story were given.

I would like to thank all of you who wrote and emailed me about the original version of the first book. Your enthusiasm and sincere appreciation and answers for more of the universe's mysteries were the motivation to write this second volume.

This second volume was written because so many more astonishing scientific discoveries were subsequently made using this revolutionary new theory of matter, space, time, energy, and the forces of nature.

In the first part of this second volume, some of the scientific discoveries from the first book are presented again [and in greater depth] because they are the foundations upon which many other great mysteries of science were explained.

Also, unlike any other book of science ever written, the visual beauty of the microscopic universe comes into view without the distraction of lengthy mathematics.

The Visual beauty of the Vortex Theory of Atomic Particles

For the first time in human history, the physics of the universe can be expressed visually instead of mathematically! In the past, the motions of the universe and the forces of the universe could only be expressed in mathematical formulas. Because most of these formulas are so complex, it took a high degree of education to understand the mechanics of the universe. *But now*, because we can see what heretofore could only be expressed in the most complicated way, children can now understand what many of their parents never could!

In honor of two of Russia's greatest scientists: Dr., Prof. Konstantin Gridnev; and Dr., Prof. Victor V. Vasiliev, I have given the vortices the name "Konsiliev Vortices".

Russell Moon

Reproduction of Prof, Dr. Victor V. Vasiliev's forward to *The End of the Concept of 'Time'*...

Shocking Commentary by Prof., Dr. Victor V. Vasiliev

Of all the millions of books that have ever been written throughout the history of mankind, only a handful have had the rare and distinct privilege of being able to change the way the entire human race forever thinks – this is one of them.

The reason for this extraordinary statement comes from the fact that this book introduces the shocking revelation that "time" – time itself – does not exist. This unprecedented discovery possesses implications of astounding proportion. Not only for all of science, but also, for all philosophy and religion. Imagine waking up and discovering every science text in the world is obsolete and every science course taught in every university in every country of the world is obsolete. NO! Except for revolutionary works of Nicholas Copernicus – *De revolutionibus orbium coelestium* – there has never been anything like this book – ever!

When Mr. Moon originally discovered this shocking and frightening revelation of many years ago, he was ignored by Western scientific community because he did not possess a PhD in Physics. Many American scientists would not even look at or read his mathematical proof! But all that has now changed. Through a remarkable series of events, Mr. Moon's discovery was introduced to the Russian scientific system: the system that possesses what can only be called: "the true scientific attitude".

The Russian scientific attitude is bold and revolutionary. Contrary to the attitude of the West, in Russia, an idea is examined first, and the individual's scientific credentials are examined second: exactly opposite to what they are in the America. Using this system, when Mr. Moon's ideas were examined they were instantly recognized for the extraordinary discovery they truly are.

This theory of Russell Moon first came to my attention in 1992 while reading the abstracts of a conference he spoke at in Fort Collins, Colorado. Then Chairman of the Electro-Physics department at the *Lenin All-Russian ElectroTechnical Institute* in Moscow, I was intrigued by its combination of simplicity and profound shocking implications. I contacted Mr. Moon and obtained a copy of the original paper.

I next revealed it to colleagues and it eventually found its way into a number of other universities. After much excitement and discourse, including the publication of a pamphlet by Dr. Leonid Samuilovisch Slutskin of the "Publishing House ZNACK" of the MOSCOW POWER INSTITUTE TECHNICAL UNIVERSITY, articles on this remarkable discovery were eventually published in *Membrana*, a Russian scientific journal. These articles created much controversy.

These publications eventually drew the attention of Professor Konstantin Gridnev, one of Russian's greatest scientists. Professor Gridnev is a Theoretical Nuclear Physicist; author of books and numerous articles on nuclear physics [25 in the last 5 years alone]; Chairman of numerous world scientific conferences sponsored by Russian government; and Chairman of Nuclear Physics Department at the "Harvard of Russia" – St. Petersburg State University, in Petergof [The world-renowned university founded by Peter the Great in 1724.]

When Professor Gridnev read Mr. Moon's papers and examined the mathematics, he was so shocked by what he saw, he came to visit me in Moscow. Then in an unprecedented move, he went to America to spend two weeks with Mr. Moon to better understand this revolutionary theory and to see just what it is capable of explaining in the field of Nuclear Physics.

This trip was well worth it because he challenged Mr. Moon to use this theory to explain many of the great mysteries and conundrums of science that until this time, have had no explanation whatsoever. Although it is hard to believe, so far, over 80 of these great mysteries of science – (listed in back of this book) in Newtonian physics, Relativity, Quantum Mechanics, and Nuclear Physics – are explained: any one of which would have awarded its author a PhD in Physics.

Some of these explanations have now been published in over 30 international scientific conferences throughout the world. Many other papers are being held back from publication because of confidential research in the new phenomenon of quantum entanglement – another recently discovered scientific mystery this theory easily explains.

Because of his scientific papers, scientific articles, previous education and research, expansion of the original thesis to include the explanation for exactly how phenomenon of time itself is created; and a revolutionary scientific experiment performed at St Petersburg State University in 2005, [subsequently entitled The Photon Acceleration Effect]; Mr. Moon qualified for, and after a lengthy refereeing process, was eventually granted PhD in Nuclear Physics. The PhD was awarded in Moscow on October 19, 2005 following resolution #7 of the Dissertation Council of St. Petersburg State University of June 30, 2005. The degree called THE ACADEMIC DEGREE OF CANDIDATE OF PHYSICAL-MATHEMATICAL SCIENCES, is equivalent to the PhD degree awarded by western universities and is given by the Russian Ministry of Education. [In Russia, higher academic degrees are awarded by government and each individual is assigned a number: KT#032771.]

This PhD now gives Professor Moon's discoveries listed in this book the backing of the Powerful Russian Ministry of Education. The list of Physicists that back this work is also impressive, especially those who are either the Chairmen or Leading Engineer of their departments in their individual universities. There has never been anything like this. Some of these distinguished scientists are: Prof., Dr. Lev Semenovich Ivlev professor of Physics and Mathematics, Chairman of the Aerosol Physics Dept. at *St. Petersburg State University*, in Petergof [author of 17 books, and 70 scientific papers]; Prof., Dr. Vladimir N. Puchinskiy of the *Moscow State Technical University* (MAMI); Dr. Alexander E. Filippov Leading Engineer of the V.I. *Lenin All-Russian Electro Technical Institute* (VIE); Dr. Alexander N. Vasil'iev Leading Engineer *of the Moscow Power Energy Institute (Technical – University)* (MEI); Prof., Nikolay Yur'evich Terekhin, Leading engineer Dept. of Atmospheric Physics, *St. Petersburg State University*, Petergof; Dr. Stanislaw I. Gusev, Chief of department, at the V.I *Lenin All-Russian Electro Technical Institute* (VIE). And of course myself - Prof., Dr. Victor Vasiliev, Chairman of the Electro-Physics department at the *Lenin All-Russian Electro Technical Institute* in Moscow; and Prof., Dr. Konstantin Gridnev Chairman of the Nuclear Physics Dept. at *St. Petersburg State University* in Petergof.

Prof., Dr. Victor V. Vasiliev; Jan. 2009

PART I
THE TRUE VISION OF THE UNIVERSE

Chapter 1
The Mystery Amid the Foundations of Science

> The unexplained mysteries of the universe sent me upon a life-long quest to discover the true vision of the universe.

Many years ago, when I was a student studying physics in college, I began to realize that a great mistake of enormous proportions existed in the 20th Century scientific vision of the universe.

I first encountered the existence of this massive problem while I was taking the first two courses in physics and chemistry. Here, I was introduced to the shocking fact that the most basic and fundamental principles of physical science that are used to explain everything else in the universe are themselves mysteries!

Although this shocking fact seemed unbelievable, it was absolutely true. *The fundamental principles of physical science* – that are also used in engineering, physics, chemistry, thermodynamics, and astronomy (and everything else we encounter in the entire physical universe) – are *"fundamental unknowns"*!

The first and most important of these "fundamental unknowns" is mass, the concept of mass itself. For those who have forgotten or for those who never knew, mass is the single most important scientific characteristic of matter. It is the one characteristic of matter that is used in all of the formulas of physics and engineering to explain the motions of matter. And yet, it is a complete mystery!

> Note: although the mystery of mass was supposedly cleared up by the "discovery" of the Higgs Boson Particle, it will be revealed in this book that – "most likely" – scientists at CERN made a mistake in 2012. Because there is no Higgs Boson Particle! The real cause of mass will be revealed in this book.

It is a mystery because although science knows what mass *does,* it does not know what mass *is!* If you draw a picture of one of the principle pieces of matter such as a proton, electron, or neutron, no scientist in the world can point to its mass. Especially the electron:

> Note: the electron is a point particle possessing no internal structure. Yet it contains mass. Supposedly, the electron's mass is explained by the "Higgs Field". [This is another false idea that will also be revealed later.] For example, ask yourself this question: all fields, such as the magnetic, electric, and gravitational fields decay with distance [$1/d^2$], why doesn't this one? Also, all fields have a point of origin – so where is the point origin of the Higgs Field?

Nor has matter's resistance to acceleration, *inertia*, ever been explained.

Mass attracts mass and mass resists acceleration. The attraction [gravity] is supposedly explained by the existence of a particle called the *graviton,* and yet, amid the trillions of collisions in particle accelerators, no scientist has ever seen one!

I have never heard anyone address the conflict created by the observation that mass attracts mass, creating movement, yet resists movement when a force tries to accelerate it. Why motion is created in one instance yet resisted in the next is not explained or defined even in the definition of mass. Mass is defined as, "an *inherent property* of matter that is a measure of the amount of matter present in a body"; yet what this inherent property of matter is, why it exists is [until now] unknown.

Gravity is a problem too. Hundreds of years ago, Sir Isaac Newton discovered the famous mathematical relationship that describes the attraction of one mass for another mass. He called it the Law of Gravity. And even though he used the principle of mass to discover this great law of physics, not even this greatest of all scientists could explain why mass attracts mass! The Law of Gravity only explains what happens, it cannot explain why it happens.

Even today, in this age of space travel and electronic internet wizardry, for scientists and engineers, the mystery of the most obvious of all the forces in the universe remains a mystery. But not anymore! Using the principles of the Vortex Theory, gravity is so easily explained, that grammar school children will one day know how to *draw* it!

Then we come to the great Albert Einstein. Just his name exudes visions of genius. We all know that Albert Einstein discovered the relationship between mass and energy and used it in his famous equation. But not even the great Albert Einstein knew what mass was. Nor do most of us know that according to Russian Scientists Einstein plagiarized this greatest of all formulas…

Note: actually, an Italian physicist named Olinto De Pretto discovered $E = MC^2$ two years before Einstein but never got credit for it? Even though it was published in a scientific journal! Why is this?

Mass is used in almost every equation of major importance in physics, engineering, and astronomy. Mass is used in the formulas that are used to explain force, momentum, inertia, acceleration, energy, work, kinetic energy, power, torque, and many more. In fact, practically every essential equation used to explain the workings of the universe uses mass. This great irony of science is most disturbing when we realize that the most well educated and pragmatic men of this era are using a total and complete mystery of the universe in their attempt to explain the other mysteries of the universe! Such as energy!

The mysteries of the universe continue to increase dramatically when we realize that *energy, the opposite of mass,* is also a mystery!

It is inconceivable to realize that we – who live in the super energy era of space travel, atomic submarines, and five hundred million cars powered by gasoline – have no more idea what energy is than a cave man sitting on a rock, staring into the flames of his campfire. Even the electric company that sends us a bill stating we have used so much power in the last month (and are charging us for it!) has no idea what they have just sold us!

Not even Albert Einstein – who won the Nobel Prize for explaining that energy travels in packets called photons causing the Photo-electric effect – knew what a photon was! Neither he nor anyone else (until now) knew how a photon was constructed. It is another of the great mysteries of 20th Century science.

Although a photon is described as an object that acts as both a particle and a wave, nobody knows why. Nobody knows why because nobody knows what is inside a photon. To understand the great significance of the failure to explain what is inside a photon, consider the following:

If photons were some sort of insignificant mundane particle, the failure to explain how a photon is constructed wouldn't matter. But the photon is extremely important. Photons are energy: heat is photons; light is photons. We cannot see without photons. We cannot read this page without photons.

Photons are also vitally important to life. We would not be alive without photons. They animate the matter within our bodies allowing us to move. Without motion we cannot survive. No creature would be able to survive without photons; in fact, life itself would not be possible without photons. They are the life of the universe, which makes the failure to know what they are a great failure indeed!

Unfortunately, even though the photons of energy are a mystery, like mass, they are nevertheless used to explain other mysteries. Especially those encountered in the science of thermodynamics!

Listen to this one: after learning that neither mass nor energy has any explanation, the following definition of *specific energy* from the science of thermodynamics seems almost like a "Stand-up Comedian's" joke. Believe it or not, specific energy is defined as *the energy per unit of mass!* [One unknown concept is being used to explain another unknown concept!]

But this is not a funny joke because specific energy is used extensively throughout the serious science of thermodynamics. Unfortunately, so is *entropy.*

Entropy is of special importance because one day in college, the failure of a professor to explain why it was occurring, shocked the hell out of me and reinvigorated my thinking about finding the massive mistake that had to exist somewhere in science. Before going to college, I do not remember ever hearing the word "entropy". But after the professor's failure to explain why it occurs, I have never forgotten it. It created one of those pivotal moments in life you never forget.

I remember sitting in the big lecture hall used for teaching Freshman Chemistry at California State University at Long Beach, when the professor wrote the word *entropy* on the chalkboard and then turned to look at us. He stated that entropy is the amount of energy that a system has used and will never be able to use again [such as the energy pouring out of the Sun]. He said, "Because of this fact, the entropy of the universe is increasing". And then he added, "Why is it increasing…" he shrugged his shoulders as he said, "…Nobody knows?"

Nobody knows? "What the Hell?" Entropy is one of the most significant observations ever made by science and nobody knows why it is occurring! Nobody knows why energy is pouring out of the Sun and all of the stars in the sky; nobody knows why heat is flowing out of all of the living creatures of the earth; or for that matter, why all of the physical matter in the universe appears to be cooling off! [Note, the Vortex Theory also and easily explains entropy.]

Although I think I was just as stunned as everyone else in the room, what impressed me was the fact that *he* did not know. He just did not know! This educated man, this distinguished professor of chemistry who was about to teach us all about this massive phenomenon occurring everywhere in the universe, and test us on it, had absolutely no idea what was causing it!

Luckily, he was also very perceptive. He was aware of the turmoil created within a student when the explanation of such an important and major concept is left unresolved. Because he next stated that this is only one of *many* unexplainable observations in science. That there are many, many more observations without explanations! There are laws that exist that we have to accept – even though we don't know why they exist!

Perhaps he said this because he knew what was coming next. A few days later, while sitting in my physics class, the professor of physics introduced us to the fundamental principles of the science of mechanics – Newton's Three Laws of Motion.

Newton's three laws of motion are absolutely essential to both the physics and the engineering sciences. In Cowles Encyclopedia of Science Industry and Technology, the importance of these three laws of motion is well stated: *"The relationship of force to motion was described in three laws formulated by Sir Isaac Newton. Without these fundamental laws the science of mechanics might well be impossible".* Most impressive!

A most impressive statement indeed! But what is even more impressive is the shocking fact that these three laws that have been with us for over three hundred years, and are the foundation for the engineering sciences; that were and are being used to design every building, ship, car, plane, and train upon this planet, are nothing more than observations without explanations! Newton did not know why they existed. Neither does any scientist or engineer know why they exist! And yet, the entire structure of our world is built upon them! [Again, now they are easily understood.]

And then…if all of this isn't enough, we come to the problem of electrostatic charges. There is something called Quantum Chromodynamics. It is a mouthful of a word and is used to explain all of the charges particles acquire during particle collisions in massive particle colliders such as the one used at CERN. Quantum Chromodynamics is a wonderful theory explaining charges for all types of exotic particles most people have never heard of such as Pions, Muons, Tau's, Mesons, Baryons, etc. A wonderful theory EXCEPT, that it possesses one great fault: it cannot explain what is called the *Fundamental Charge of Nature*. It cannot explain the charge on the one little particle we all know the name of – the electron! That's right! The one particle we all know about is a mystery to all of these Quantum Chromodynamic experts!

For example, just casually ask one of these highly educated PhD physicists this question, "What is creating the electrostatic charge on the proton?" And, in a very learned and distinguished tone of voice, this is the answer you will get, "The charge on the proton is being generated by three quarks within the interior of the particle." But then you say, "Wow, isn't that impressive!" And he will look at you like he is an educated and impressive individual. But then you ask him this next question: "Well, if that is true, then how come the tiny little electron that has no quarks or no internal structure but possesses the same but opposite charge as the much larger and massive proton with all of its quarks?" And you will hear silence! Because there will be nothing he can say! Because he has no answer!

Not one of these most distinguished tenured professors in the most highly respected universities in the world can answer this seemingly trivial question about the little particle all of us know about; the particle used in all of the electronic components in all of the electronic equipment in the world! Not one of them! [But you can. You will be able to answer this question after reading this book: because only the Vortex Theory can answer this question!]

Well so what? What does it matter? What if there are problems with the fundamental principles of science and engineering that no scientist or engineer can answer? Has it hurt us? Has it had any effect on the great astronomical discoveries we have made about the universe, or in the wonderful electronic technologies we have developed?

The above question is very important because the answer to it is even more important. It goes like this: if everything we have today was achieved through ignorance and error, just imagine what we can accomplish using the truth!

Chapter 2
The Problem Is Time!

> So where is the problem? And even more important, how are we supposed to solve it?
>
> If Sir Isaac Newton and Albert Einstein, the icons of science, were unable to explain the most basic and fundamental principles of science, what chance do the rest of us have? If the best of the best have all failed, how can an ordinary person do any better?

Well, there is indeed a way to succeed where others have failed, and actually, it's quite simple. It goes like this; if we can find out why others have failed, we can avoid doing what they did wrong. By avoiding what they did wrong, we can avoid their mistakes. And by avoiding their mistakes, we at least have a chance to succeed; and sometimes, that's all we need.

So why did these two great men fail?

After many years of research, it is now easy to answer this question. They failed because they *believed* like their predecessors believed! Then, because they believed like their predecessors, they thought and reasoned like their predecessors. [And we shall soon see that even though Einstein thought he was thinking differently from Newton, he wasn't.]

In reasoning like their predecessors, they came to the same conclusions as their predecessors. In doing so, they failed to realize that when everyone thinks alike, everyone reasons alike, and everyone comes to the same conclusions. Passing on errors from one generation to the next. The only way to break the cycle, and to reach new conclusions nobody has ever reached before, is for one to think in ways that nobody has ever thought of before.

The idea that the world was the center of the universe was reaffirmed from one generation to the next for countless generations. All anyone had to do was look at the sky and observe the motions of the Sun, Moon, planets, and stars. Since it was obvious for everyone to see that everything appeared to be revolving around the Earth, it was easy for each new generation to come to the same conclusion as the old generation – the Earth had to be at the center of the universe. And this belief might still be with us if it weren't for one man – Copernicus.

Copernicus successfully broke free of the thinking and hence the reasoning patterns of the previous generation. He thought differently from everyone else. And thinking differently he reasoned differently from everyone else. And when he did, he changed both the thinking and the reasoning of the entire world! A simply awesome accomplishment!

Columbus did it too. It is awesome to contemplate that again, one bold and audacious man thinking differently from the rest of the world, changed the way all of us think and reason today.

Columbus believed the world was round.

> Note: Columbus was not the first. Twenty-five hundred years ago, a near-contemporary of Aristotle in Athens named Heraclides, who was laughed at by the people of his era, tried to tell others he believed the Earth was huge, was round, and was rotating upon its axis.

In doing so he both thought and reasoned differently from the people of his era. His thinking must have seemed not only radical but also preposterous. He came up with the seemingly outrageous idea that to go east, all one had to do was go west! This was a simply extraordinary conclusion that nobody from his flat earth era had ever thought of before! A conclusion that was absolutely true yet missed by those of past generations because their erroneous beliefs created erroneous thinking that created erroneous reasoning. A situation unbeknownst to the world that exists right now!

Although it is hard to believe, everyone in the world today is just as ignorant about the construction of the physical universe as Columbus' generation was about the construction of the Earth. The mistake comes from the fact that an ancient error in the "non-scientific vision of the universe" was used by Newton in his scientific vision of the universe and was passed on to all future scientists including Einstein. And this error is still causing everyone to continue thinking and reasoning in the same erroneous way they did!

So, what is this ancient error?

This error is time. Time - itself!

Today this error is so obvious it seems simplistic, but many years ago the problem of just trying to decide where to start looking was overwhelming. There was such an enormous amount of material to consider, with no idea what to look for. But after much consideration, it was finally realized there was a way to simplify everything. Everything that exists in the universe could be placed into one of just five categories: matter, space, time, energy, and the forces of nature.

The categorization was the easy part. The contemplation of these five parts took many years. But eventually, the effort was rewarded, as the concept of time slowly became suspect.

What made time suspect comes from the fact that it is *an ancient concept of unknown origin*! Time is so old it predates the discovery of writing. Nobody knows who developed the concept of time!

Because of this fact, it was further realized that the people who developed the concept of time were ancient peoples who were totally unaware of the astronomical motions associated with time.

This lack of scientific knowledge is very important. Was dawn age man really sophisticated enough to recognize such an abstract concept as time?

The concept of time is a very abstract idea. It cannot be perceived with the senses. It can only be deduced through observations associated with motion. This observation is extremely important, for many of our conclusions are based upon our perceptions, but sometimes our perceptions are based upon our conceptions.

Consequently, was dawn age man cognizant enough of his mental processes to differentiate between the two? [Is anybody?] Spoken more bluntly, did they perceive time, or did they conceive time? Did they invent the concept of time in an effort to explain what they did not understand?

When considering this hypothesis, we tend to forget that the concept of time *is not* a concept that we as 20th century men developed. It is a mental legacy inherited from early man and passed on from generation to generation for thousands of years. A concept developed within the reasoning process of the child like superstitious minds of our ancient ancestors - men that were totally ignorant of even the most simple and basic astronomical knowledge we take for granted today.

Today, it is hard for many of us to accept the fact that grammar school children know more about the motions of the Earth and the solar system than the greatest philosophers in the Golden age of Greece. It is hard to believe that six and seven-year-old kids know more about the orbits of the planets than Plato, Socrates, and Aristotle!

Because the motions of the Earth, Moon, Sun, and planets are so obvious to everyone, it is hard for us to envision a time when men knew nothing about even the most rudimentary motion of the Earth. It is also easy for us to forget today that ancient peoples did not know the Earth was rotating, creating the effect of night and day – and creating the effect that everything was circling it. Or that the changing location of the Earth in its orbit around the Sun was responsible for making different star constellations appear at different seasons of the year; or that during its orbit of the Sun, the Earth's 23.5degree tilt was creating the four seasons.

In fact, everything ancient peoples were observing in the sky, were effects being created by motions they were totally unaware of!

It gets even more uncomfortable when we realize our concept of time was developed by men who worshipped fire, feared thunder and lightning as the wrath of the gods, and made animal sacrifices for good crops and successful hunts.

In all honesty, I must admit that when I first thought these thoughts, I found myself wondering if anyone else in history had ever dared to think like this before? The concept of time is so much a part of our daily lives that to question its validity, even within the privacy of our own thoughts seems ludicrous – if not insanity itself.

But then, "just because" everyone "knows" time exists, doesn't mean it does! Once upon a time everyone "knew" the world was flat, everyone "knew" the Earth was at the center of the universe, and everyone "knew" that Aristotle's vision of an unchanging universe was absolutely correct. Once, to be a member of the educated elite, everyone studied Aristotle and knew that the stars were in stationary positions in the night sky. But what is most disconcerting of all is the fact that all of these totally false and erroneous ideas endured from generation to generation for almost two thousand years!

Man has grown comfortable with his belief in time. But suddenly I wasn't. I began to suspect that the concept of time might be nothing less than a total and complete mistake! Perhaps the greatest blunder mankind has ever made! But is there any way to confirm this suspicion?

In seeking to answer this question it was realized that any idea that has to be modified when new discoveries are made is probably wrong. And this modification has already occurred with time.

Newton believed time was linear [the same everywhere in the universe]. For several hundred years this was believed to be true until Einstein decided time was relative and incorporated this new idea into his famous Theory of Relativity. Unfortunately, he then used his conception of time to explain other things.

Einstein used time to explain how space was constructed; this explanation of space then defines how matter is constructed; while the explanation of matter is then used to explain how energy and the forces of nature are constructed.

But if time does not exist as a fundamental principle of the universe, the explanation of everything based upon its existence - matter, space, energy, and the forces of nature – is suddenly suspect and needs to be re-examined. [And it has been!] However, because the worldwide belief in the concept of time has become such an unquestioned part of everyone's lives, it is necessary to reverse this error and let everyone see time for what it really is. The way to do this is with the following curiosity:

To discover the truth about time, consider the following philosophical question: if all of the motions of everything in the universe stopped and then started again, is there any way to tell for how long they were stopped?

The answer is no. Because just as a physical yardstick measures the distance between two objects, a regularly reoccurring sequence of events [harmonic motion] is used as a yardstick of time to measure the distance between a random set of events. However, unlike the physical yardstick, when the harmonic motion used to keep track of time stops, the yardstick of time no longer exists and the distance between random events can no longer be measured. Hence, we are unable to tell if the universe was stopped for a second, a day, a year or a million years.

Consequently, we are faced with a dilemma: if time exists as a fundamental principle of the universe, how come it ceases to exist when motion ceases to exist?

The answer could be that time is not a fundamental principle of the universe. If time only exists because motion exists, then *time is a function of motion, a phenomenon created by motion – a hypothesis possessing profound implications!*

The concept of time plays a major part in both the explanation of how the universe works, and how the universe is constructed. Not only is the concept of time a part of almost every important physics equation used to explain the fundamental properties of the universe, it is also viewed as one of the five basic parts of the universe: matter, space, time, energy, and the forces of nature.

But as we have seen before, because all five parts interact with each other, the construction of each one of the five parts depends upon the construction of the other four. Eliminate one part and the constructions of the other four parts have to be completely revised.

So, what happens to the other four when time is eliminated? Or expressed in another way, what happens to the 20th Century scientific vision of matter, space, energy, and the forces of nature when time is eliminated as a fundamental principle of the universe?

The answer is startling! Since all of the other theories explaining the universe's construction depend upon the existence of time, if time does not exist, *every one* of them becomes instantly obsolete! It forces us to completely rethink the universe's construction; and look for a solution nobody ever thought of before.

Chapter 3
The Undiscovered Territory

> Amongst the billions of people of the earth thinking billions of thoughts every day, to think of something nobody has ever thought of before is the *Undiscovered Territory*. It lies in a realm unique unto itself. And this is exactly what the true vision of the universe is.

The true vision of the universe is unlike anything anyone has ever seen or imagined before.

Man's vision of the universe is based upon five fundamental principles: Matter, Space, Time, Energy, and the Forces of Nature. Everything in the universe is constituted of or constructed out of one or more of these five pieces. Our beliefs in how they act and interact are responsible for the scientific equations we presently use to explain everything we see. Our knowledge of physics, chemistry, and all of the physical sciences is based upon these five pieces. Our technology and our engineering strategies are based upon these five parts. Even our philosophy, religion, and personal self-images as human beings are based upon or influenced by our understanding of how these fundamental pieces are constructed. In fact, the more we think about it; almost every aspect of our lives is somehow influenced by this vision.

Destroy this vision and everything as we know it – disappears with it. But the cost is worth it. Because the vision that reappears in its place is unlike anything anyone has ever imagined before!

The discovery of this vision is one of the greatest moments in the history of the world. The true vision of the universe is a fantastic vision.

The five pieces of the universe - Matter, Space, Time, Energy, and the Forces of Nature are not constructed in the way 20th Century science presently believes. They exist in a manner unlike anything anyone has ever imagined before. Like a world turned upside down, nothing is as it appears.

Matter, instead of being made of something, is made of nothing; Space, instead of being made of nothing, is made of something; Time doesn't exist; "Force" doesn't exist, while Energy and the four "Forces of Nature" are constructed out of *dense and flowing space*! In fact, *everything in the universe is made out of only one substance – space!* ["Space" is not space! It is not a void!]

Since space is the "substance" that matter, energy, and the forces of nature are all created of, it is only proper to begin our new and revolutionary vision of the universe with "space". However, since matter, energy, and the forces of nature are all interconnected, and it is impossible to describe the workings of one without describing the workings of the others, it must be understood that the explanation of each is not complete until the explanation of everything is completed.

Something else must be said too:

Chapter 4
The True Vision of Space

Contrary to popular belief, space is not a void. Space is a multi-dimensional substance that everything in the three dimensional universe is made of. This idea is not a return to the old Aether theory. As we will soon see, the space within which we exist, is unlike anything anyone has ever imagined before.

The key to discovering how *everything* in the universe is constructed is found in the construction of space.

Contrary to present beliefs, space is not a void. Space is made of something. The substance that space is made of is totally unique from our point of view. It can both stretch and flow, is constructed out of at least seven dimensions, and is in a state of expansion.

Furthermore, our three dimensional universe is not infinite. The three dimensional universe only appears to be infinite from our perspective. The three dimensional space in which we exist is in reality the finite surface of a fourth dimensional volume of space.

Just as a two dimensional plane is the surface of a three dimensional object, the three dimensional space of our universe is the surface of fourth dimensional space; and fourth dimensional space is the surface of fifth dimensional space, etc. this is not supposition, it is based upon the work of two of the greatest mathematicians who ever lived: *Joseph-Louis Lagrange*, and *René Descartes*.

Hundreds of years ago, these two great men, amusing themselves by writing numbers on pieces of paper, and discovering new formulas, never realized the implications and consequences of what these formulas actually represented! For example:

It is easy to say, and no one will disagree, that we live in a vast, three dimensional volume of space. Everyone also will agree that because we live in three dimensions, the three components of three dimensional space called length, width and height, [identified mathematically as the three co-ordinates X, Y, & Z of the Cartesian Coordinate System] are subsequently used to define and identify the position of an object. However, is it just a coincidence that a three dimensional volume of space just happens to be the surface area of a fourth dimensional sphere!

Although most people are aware of the fact that a sphere such as a basketball or a bowling ball of radius "R" has a two dimensional surface area: expressed mathematically as $4\pi R^2$; where the symbol R^2 = a two dimensional flat area. This two dimensional area encloses a three dimensional internal volume of space or matter, expressed mathematically as: $4/3\pi R^3$; where R^3 = the three dimensional volume of space.

Figure 4.1

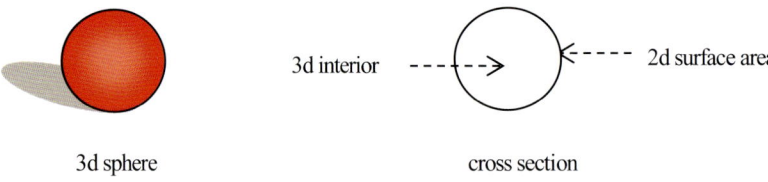

3d sphere   cross section

However, one of the great, <u>*unrealized shocking truths of the universe*</u> - discovered unwittingly by these two brilliant mathematicians hundreds of years ago - is the fact that a *three dimensional volume of* space is actually the *surface* area of a fourth dimensional object! Note: the *surface area* of a fourth dimensional sphere of radius "R" is equal to $2\pi^2R^3$: where R^3 represents the fourth dimensional sphere's <u>surface *area*</u>! This <u>*AREA*</u> is in reality a <u>*VOLUME*</u> of three dimensional space! A three dimensional volume of space surrounding an internal fourth dimensional volume equal to $1/2\pi^2R^4$. And again, it is important to realize that the <u>*surface area*</u> of the fourth dimensional sphere, is *<u>actually a three dimensional volume of space!!!</u>*

Figure 4.2

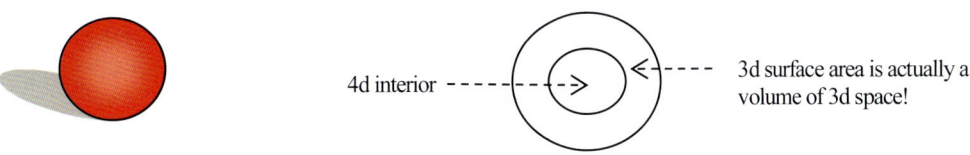

Schematic representation of a 4d sphere   Schematic representation of the cross section of a 4d sphere

What these mathematicians failed to realize is that these formulas reveal that our universe's three dimension volume of space is actually the surface area of a fourth dimensional object! This conclusion seems so outrageous that we immediately want to know if there is any other corroborating evidence to confirm this shocking conclusion? Is there?

The answer is yes, it starts like this…

As we look out into the universe with our most powerful telescopes, from our point of view, it appears we live in a vast three dimensional volume of space. However, there are problems with this vision. If everything that we see in the universe is a result of the Big Bang, where is its point of origin?

In 1054 AD, the people of the earth witnessed a supernova. Although they did not know what they were seeing, the Chinese wrote about a star that suddenly appeared and was so bright that it could be seen in the middle of the day; in the southwest, Native Americans drew stars on rocks in awe of this one of a kind event. Today we can witness the remnants of that explosion. By looking at the Crab Nebula and measuring the Doppler shift of its light, we can tell by the velocity of the particles streaming out into the universe that the nebula is the result of a gigantic explosion that occurred approximately 950-970 years ago [approximately equaling 1054 AD]. We can also use the same technique to trace the coordinates of the particles streaming out into the universe from the Crab Nebula to determine the point of origin of the supernova. Hence, we can see approximately where the blast occurred.

So, how come we cannot do this with the Galaxies of the universe? How come we cannot use this same technique to determine the point of origin of the Galaxies created after the Big Bang?

Unfortunately, this same scientific method fails when we try to apply it to the motion of the Galaxies out of which the physical universe is constructed. Even more amazing, we find it hard to believe that when we look for the origin of the Big Bang, we discover that every point in the universe appears to be the point of origin we are looking for! How can this be?

Perhaps it can be best explained using this analogy: just like an imaginary microscopic creature living upon the 2d surface of the expanding balloon, because the entire surface is expanding at once, there are no beginning co-ordinates for anything upon its surface. Also, from his point of view, as he looks outward upon the surface of the sphere and rotates through 360 degrees, from his point of view, he appears to be standing upon the center of the sphere.

The same is true for us living upon this 3d surface and searching for the beginning of the Big Bang. What we label as the Big Bang was actually the point of origin for the beginning of the expansion of the 3d surface we live upon. Also, from our point of view, just like the imaginary creature living upon the 2d surface of the 3d sphere; no matter where we are appears from our perspective to be the center of our universe. Which forces us to solve another problem: where is the center of mass of the universe?

Every second semester engineering student knows the technique used to find the "Center of Mass" of an object. It is easily accomplished using a little Physics and Calculus. It is also used to determine the center of mass of planetary systems like ours, and the center of mass of Galaxies, and Galaxy clusters. So how come this same technique doesn't work when it comes to finding the center of mass of *all* the Galaxies existing in the universe?

And again, just like the imaginary though intelligent microscopic creatures living upon the surface of the balloon, and finding that they cannot measure the mass of the balloon, because unbeknownst to them, even though the material of the balloon exists upon its surface, the center of mass of the balloon resides in the empty space within the center of the balloon: and so it is with us. Even though from our perspective, we appear to live in a three dimensional volume of space, this space is really the surface of a higher dimensional volume of space. Consequently, the center of mass of all the Galaxies that exist in the universe lies within the *center of the fourth dimensional volume* and not upon the surface! Note: this same explanation for the center of mass still works if indeed fourth dimensional matter exists within a fifth dimensional volume of space.

Is there any more corroborating evidence? Again the answer is yes; and brings us to the subject of infinity…!

Infinity is a problem most of us have tried yet failed to rationalize and reconcile within our minds. It is hard to imagine a never ending volume of space that goes on forever and ever. But that is what we see. When we look out into the universe, it seems to go on forever and ever. Luckily, it is just another illusion.

Returning again to the imaginary though intelligent microscopic creatures living upon the surface of the balloon, we find that they encounter the same mental problem: where is the end of the two dimensional surface they live upon? If the balloon is large enough, they find that if they go in search of its end, they can keep on traveling around and around it forever and ever, making them believe it is infinitely large. And this is the exact dilemma faced by us.

It appears from our point of view that the universe is never ending, that it goes on forever. But again, this problem is resolved when we realize this is merely an illusion, it is not real. If we could travel fast enough and far enough, we would come back to the exact same point we started out from! [Like Columbus, we could go west by going east!] Consequently, infinity is just another of the many illusions that are created from our limited point of view here within this three dimensional volume of the "surface area" of this fourth dimensional object!

Because many people reading this book are not scientists, a brief description of the dynamics of these higher dimensions of space is necessary. Unfortunately, because the fourth and fifth dimensions (called higher dimensional space for simplicity) are impossible to draw or describe, only analogies or inferences can be used to explain their unusual properties and qualities. It goes like this…

Roughly speaking, higher dimensional space is a volume of space located outside of our three dimensional universe. This volume of space has a unique perspective: an observer in three dimensional space cannot see anything that exists in higher dimensional space, however, from higher dimensional space, an observer can see everything that exists in three dimensional space!

The entrance into fourth dimensional space is also unique. To enter into fourth dimensional space, one must pass through three dimensional holes. Just as we must pass through two dimensional holes in three dimensional volume of space, we must pass through three dimensional holes to enter into the fourth dimensional volume of space (and fourth dimensional holes to enter fifth-dimensional space, etc.). Revealing that A Hole is Always One Dimension Less than the Number of Dimensions Present!

The relationship between the numbers of dimensions a hole is made of can be used to determine how many dimensions of space exist in the universe. Because a surface is always one dimension smaller than the volume it encloses, the entrance or hole leading into the interior of a volume is always one dimension smaller than the number of dimensions a volume is constructed out of.

For example, a characteristic indicating the existence of three dimensional space is found in the presence of two dimensional holes. It is easy to see that all holes in three dimensional space are two dimensional. All doors, windows, cave openings, pipe openings, and tunnel openings etc. are two dimensional holes.

The same is true for the existence of fourth dimensional space. If fourth dimensional space exists, a characteristic of its presence will be the existence of three dimensional holes somewhere within our physical universe. [We will find these holes in the next chapter.]

Unfortunately, great difficulty is encountered when first trying to visualize three dimensional holes and higher dimensions of space. This problem occurs because we both think and reason in terms of three dimensional space. Whereas, to think and reason in terms of fourth dimensional space, or fifth-dimensional space, we have to visualize the universe in a way that is totally different from anything we have ever done before.

A way to envision the relationship between the different dimensions of space is found in the following drawings:

Figure 4.3

One dimensional space consists of one line.

Figure 4.4

Two dimensional space is a plane that is at right angles to the line.

Figure 4.5

Three dimensional space is at right angles to the plane.

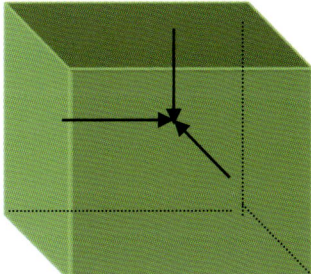

Figure 4.6

Fourth dimensional space is at right angles to three dimensional space. Since fourth dimensional space is impossible to draw, the following is only a representation, a schematic drawing.

Another way to understand this construction is to look up at a corner of the room you are sitting in. Observe the point where the ceiling and the two walls meet. The ceiling and the two walls represent the three dimensions of space we live in: length, width, and height. Now observe how the ceiling is at right angles to both walls, while both walls are at right angles to the ceiling and the opposite wall. This relationship allows us to observe the fact that each dimension is at right angles to the other two dimensions *simultaneously*.

Although the ninety-degree angles that exist between the three dimensions and create three dimensional space are easy to see, the ninety-degree angle that exists between fourth dimensional space and the three lower dimensions is impossible to visualize.

Even though we cannot visualize the ninety-degree angle that exists between three dimensional space and fourth dimensional space, what we can do is to visualize what the entrance into fourth dimensional space will look like from our three dimensional perspective. Because this entrance is a three dimensional hole, it will be spherical in shape; and if it is small - very small - it will appear to be a tiny spherical particle!

Figure 4.7

Chapter 5
The Proton and the Electron

The present viewpoint of the universe states that protons, electrons, and neutrons are particles. This is a mistake. Protons, electrons, and neutrons are not particles. Protons, electrons, and neutrons are three dimensional holes existing upon the surface of fourth dimensional space. Three dimensional space flows into and out of these holes creating the electrostatic force of nature.

The truth about matter is absolutely shocking. The matter of the universe is not "matter!" Protons, electrons, and neutrons are not made of "something". Instead, they are tiny three dimensional holes in space. All of the matter that exists everywhere in the universe – including the matter we are made out of – is made out of nothing at all! The stars, the Sun, the Earth, the planets, moons, asteroids, comets, gas clouds, and the bodies of all living creatures including our own physical bodies are nothing more than vast collections of incredibly tiny three dimensional holes existing upon the surface of fourth dimensional space.

Because everything is made out of protons, electrons, and neutrons, it is these holes that will be discussed first - bringing us to the proton…

The proton is a three dimensional hole bent into the surface of fourth dimensional space. Three dimensional space flows into the proton creating its electrostatic charge.

[Making a rough analogy, the proton can be compared to the drain at the bottom of a sink that water is flowing into.]

Figure 5.1

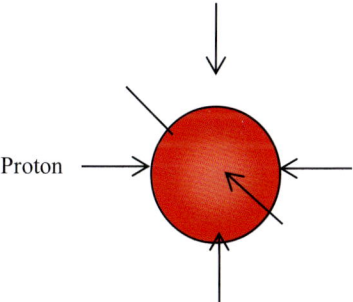

Although 20th Century science shows the lines of electrostatic force pointing out of the proton, this is a mistake. It was Benjamin Franklin who originally and arbitrarily assigned the lines of electrostatic force. Neither Franklin nor anyone else knew which way they pointed until now. The clue to which way they point is found in the gravitational field that surrounds stars and will be discussed in the section dealing with gravity.

The proton is also rotating around a fourth dimensional axis [see PART II #4]. This rotation is responsible for its ½ spin, and its magnetic moment.

Note: the proton's quark content will be discussed in Book 3 *The Explanation of the Quark Theory.*

When a proton is formed, three dimensional space is pulled into higher dimensional space causing three dimensional space to flow into the hole. As the three dimensional space surrounding the hole is pulled into the hole, a sphere of *less dense* space surrounding the hole is created:

Figure 5.2

Note: in relation to the size of the proton, the region of less dense space that surrounds the proton is so massive it is impossible to draw the proportionate sizes. However, if the proton were the size of a grain of sand, the region of less dense space would be the size of a football stadium

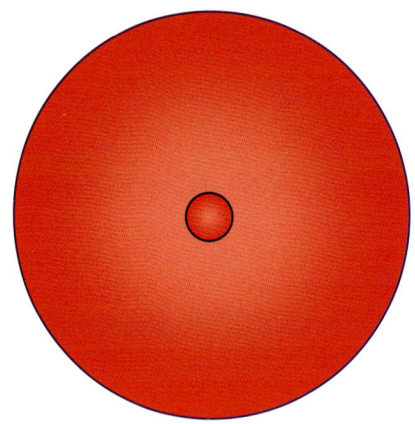

.

The less dense region of space that surrounds the proton is a function of the volume of three dimensional space the hole is traveling through. As the hole moves through the three dimensional space of our universe, the space the hole is passing through bends inward towards the hole as it approaches [creating the less dense region] and then back outward [returning to normal density] as it passes by. Consequently, even though this massive sphere of less dense space constantly surrounds the proton, it is the *region of space the hole is passing through* that is creating this sphere. The same is true for the electron but in the opposite direction:

Because of this relationship, it can now be revealed that the particle aspects of the proton are created by the hole, which its wave effects are created by surrounding space.

The electron is revealed to be a three dimensional hole bent out of the surface of fourth dimensional space. Three dimensional space flows out of the electron, creating its electrostatic charge.

[Making another rough analogy, the electron can be compared to a tap, or faucet, that water is pouring out of.]

Figure 5.3

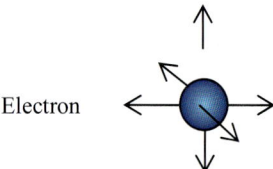

Just as the proton is a three dimensional hole bent into fourth dimensional space, the electron is a three dimensional hole bent outward, out of fourth dimensional space. Just like the proton, the electron is also rotating around a fourth dimensional axis of rotation [see PART II #4]. This axis of rotation is responsible for creating the electrons ½ spin and its magnetic moment.

Later, it will be shown how the electron is really the other end of a whirling vortex of three dimensional space. This whirling vortex of three dimensional space flows into the proton and then out of the proton into higher dimensional space; after it passes through higher dimensional space, it then returns back into three dimensional space via the electron.

As this vortex of whirling space flows out of the electron, it pushes the space surrounding the electron outward, surrounding the electron with a massive region of *dense* space. Although this region is massive in relation to the size of the electron, it is much smaller than the size of the less dense region of space surrounding the proton.

Figure 5.4

And as with the proton, because the region of dense space surrounding the electron is so incredibly massive, it is impossible to draw its correct proportions.

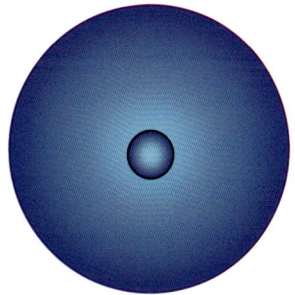

And again, just like the region of less dense space that surrounds the proton, the region of dense space that surrounds the electron is a function of the three dimensional space the hole is traveling through. This means that as the hole moves through the three dimensional space of our universe, the space the hole is passing through bends outwards away from the hole as it approaches, and then back inward as it passes by. Consequently, even though this massive sphere of dense space constantly surrounds the electron and seems to travel with it, the region of space the hole is passing through is creating this sphere.

> Because of this relationship, it can now be revealed that the particle aspects of the electron are created by the hole, while its wave effects are created by the space surrounding it.

Because there is presently no units or nomenclature with which to describe the density of space, the densities of the space surrounding the proton and the electron can be expressed in terms of lines per centimeter.

Figure 5.5

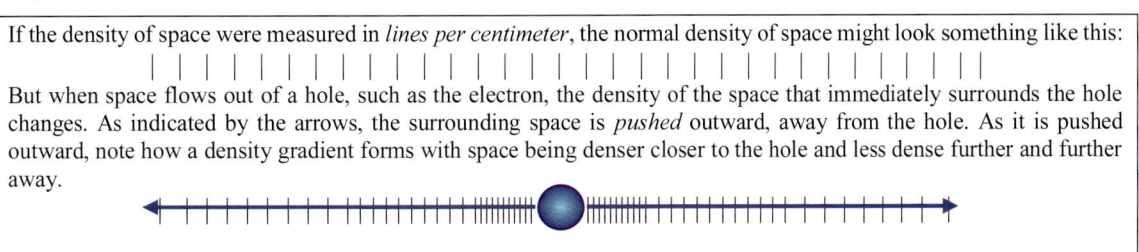

The opposite would be true for space flowing into a hole. If three dimensional space were pulled into a hole, the space in the region directly surrounding it would be stretched into the hole. This stretch would create a spherical region of less dense space immediately surrounding the hole, a region that slowly begins to return to its normal density further away from the hole.

Figure 5.6

The schematic representation below shows a density gradient surrounding a hole (such as a proton) that *space is flowing into*:

As indicated by the arrows, the surrounding space is *pulled* into this hole. Note too, how the gradient formed here is less dense closer to the hole and denser further and further away from the hole.

Note: this same region of bent space also surrounds the neutron. The creation of the neutrons less dense region will be discussed in the section describing the neutron.

The knowledge of the hole in space surrounded by the region of dense or less dense space finally allows us to explain the mystery of the PARTICLE AND WAVE THEORY OF MATTER: See PART II #2

Chapter 6
The Vortex [Part 1]

*In higher dimensional space, the protons and electrons **in atoms** are connected by a flowing vortex of three dimensional space. This vortex begins as space flows into the proton, then out of it and into fourth dimensional space, then through fourth dimensional space into the electron where it again emerges into three dimensional space.*

The proton and the electron are connected by a vortex of three dimensional space flowing from the proton to the electron in fourth dimensional space!

Figure 6.1

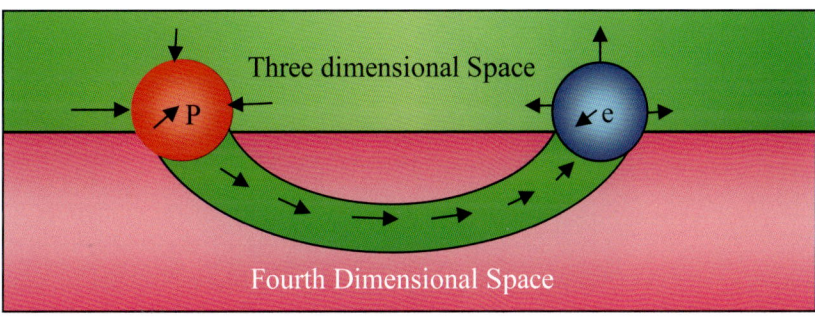

When a hydrogen atom is created, although present day science says that the electrostatic charges of protons and electrons pull these two "particles" together, this is an oversimplification. What really happens is that some of the space flowing out of the electron begins to flow into the proton; as these two holes move closer together, a critical distance is reached where all of the three dimensional space flowing out of the electron flows directly into the proton. When this situation occurs, *a second* vortex of whirling space is created.

The second vortex now flows from the electron back to the proton in three dimensional space.

These two vortices create a circulating flow containing a fixed volume of space.

This circulating volume of three dimensional space continually flows from the proton, into fourth dimensional space - through higher dimensional space, and then into the electron. Here, it exits the electron, flowing back through three dimensional space and into the proton once again, binding the proton to the electron creating a hydrogen atom.

Figure 6.2

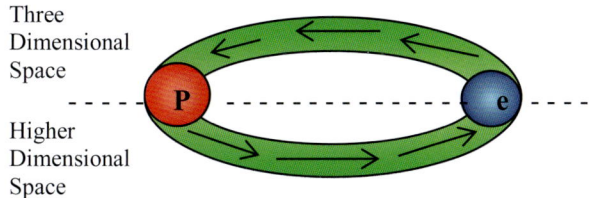

When the circulating flow commences, both of the electrostatic charges are neutralized. The word "neutralized" was used because no flowing space escapes from the system. If surrounding space still flowed into or out of this system, atoms would possess electrical charges (and every time we touched something we would get shocked).

Because these vortices exist in all atoms throughout the universe, the revolutionary new vision of matter presented within this book was christened *The Vortex Theory of Atomic Particles*.

Because the diameter of the proton appears to be about one thousand times larger than the diameter of the electron, the three dimensional vortex is shaped like a cone!

Figure 6.3

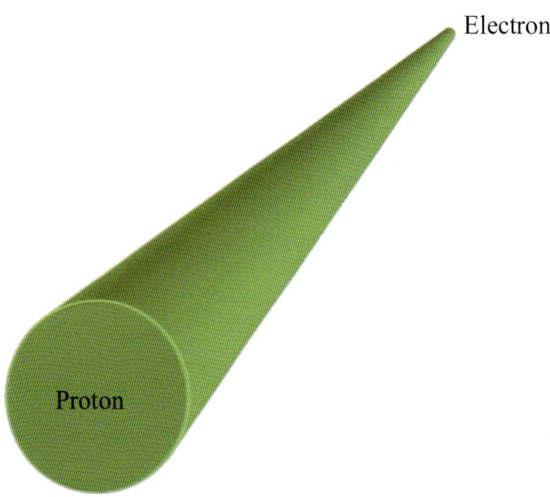

But the cone shape is deceiving. Even though it appears as if the volume of space flowing into the proton is more than the volume of space flowing out of the electron, this is an illusion.

Because the electrostatic charge of the proton and the electron are the same, the volume of three dimensional space flowing into the proton and out of the electron are equal. This means that the volume of the three dimensional space flowing out of the "surface area" of the electron is exactly the same as the volume of the three dimensional space flowing into the "surface area" of the proton. Since these surface areas are of different sizes, but still possess the same volume of three dimensional space flowing through them, the density of the vortex is different at each end.

Because the proton is approximately one thousand times larger than the electron, the surface area of the proton is about one million times larger than the surface area of the electron. This relationship reveals that the density of the vortex at the electron is at least one million times denser than the density at the other end of the vortex that enters the proton.

The size of the proton and the electron are also a clue to the density of the space immediately surrounding them. When the proton and the electron are not connected to each other in an atom by the three dimensional vortex, but instead, exist apart in three dimensional space, the difference between the densities of the space at their surfaces is at least one million times greater at the electron than at the proton. This observation will become very important later on, when the fifth force in nature is discussed.

It should also be mentioned that since the three dimensional volume of the vortex flows into and out of each sphere, the surface area of either sphere is actually the cross-sectional area of the vortex at that point. Furthermore, because the ends of the vortex are of different sizes, the cross sectional area of the vortex that was used in the mathematical proof of this theory was the cross sectional area at the halfway point between the proton and the electron in a hydrogen atom. The construction of the higher dimensional vortex is most unique.

Because the higher dimensional vortex is not a volume but instead, a surface of fourth dimensional space, it is impossible to draw. However, even though it is a surface, its fourth dimensional "surface area" is equal to the three dimensional volume of the three dimensional vortex in three dimensional space.

This relationship between the three dimensional hole and the fourth dimensional volume can be demonstrated in this unique two-dimension to three-dimension analogy:

Figure 6.4

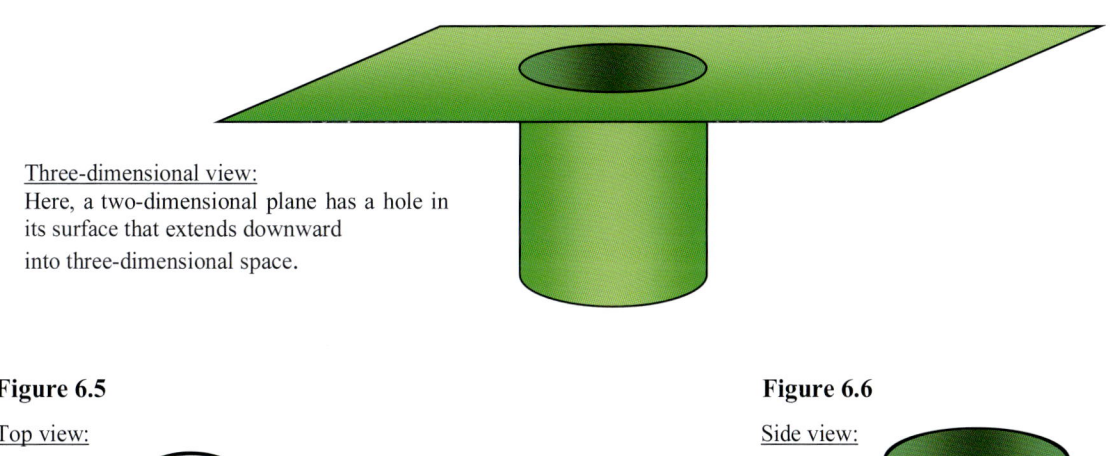

Three-dimensional view:
Here, a two-dimensional plane has a hole in its surface that extends downward into three-dimensional space.

Figure 6.5   **Figure 6.6**

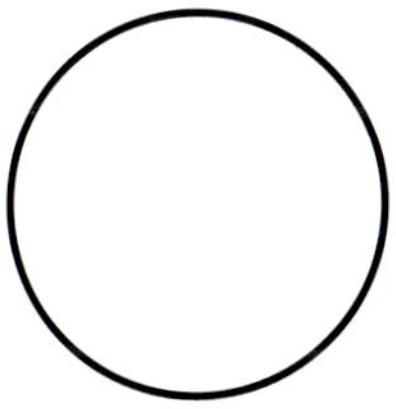

Top view: Side view:

The above three drawings share an unlikely continuity. In Figure 6.4, if the two dimensional green surface of the plane were flowing down into three dimensional space via the circular hole, the two dimensional "volume" of the flow would be equal to:

$$V = (\pi d)(v)$$ Where: $\pi = 3.14$
d = diameter of the circle in meters (m)
v = velocity of two dimensional flow in m/sec
V = volume of the two dimensional flow in m^2/sec

Consequently, the volume of the flow would be equal to the length of the circumference of the circle multiplied by the velocity of the two dimensional surface that is flowing downward around the edge of the circular hole.

As this two dimensional surface flows downward, it forms the surface of the three dimensional cone. This flow creates a unique continuity between all three shapes: even though the cone forms a three dimensional shape, the two dimensional surface area of the cone is equal to the two dimensional "volume" that flowed into the circumference of the hole and created the cone shape.

21

Applying this analogy to three and four dimensional space, we can see how the three dimensional volume of space - that flows through the two dimensional surface area of the three dimensional hole and into higher dimensional space - creates a fourth dimensional shape whose three dimensional surface area is equal to the volume of the three dimensional space flowing through the hole.

This relationship can be expressed in the following formula:

$$V = (4\pi r^2)(v)$$

Where: $\pi = 3.14$
r = radius of spherical hole in meters (m)
v = velocity of the three-dimensional in m/sec
V = volume of the three-dimensional flow in m3/sec

Consequently, even though the three dimensional vortex is a volume and the fourth dimensional vortex is a surface, the same amount of three dimensional space is flowing in each vortex!

However, the next big question regarding proton-electron pairs is how they became *entangled*?

Particle collisions in linear accelerators reveal that when a proton is created, an anti-proton is created with it; and when an electron is created, its anti-particle [the positron] is also created with it too. So how do protons and electrons become entangled creating hydrogen atoms?

The above question is answered by first revealing that the northern lights reveal that electrons and other negatively charged "particles" coming from the sun are attracted to the north pole of the earth [the south pole of a magnet]. This reveals that the vortices between electrons and positrons can exist in long filaments without breaking up. However, what happens if two higher dimensional vortices cross each other in higher dimensional space?

From experimentation with electronic circuits, it is known that electric current always seeks the shortest path between two different charges. Therefore, it is not unreasonable to assume that if the motions of two holes in three dimensional space caused their vortices to cross each other in higher dimensional space, the two vortices would break and reconnect themselves to the particular hole completing the shortest path back to the three dimensional surface.

The mechanism via which this switch would occur would be the change in the cross sectional area of the vortex. Because the electron is smaller in size than the proton, the cross-sectional area of the vortex decreases as it approaches the electron, increasing its density. *This increase in density would be seen by a longer vortex with a larger less dense cross-sectional area as an increase in the electrostatic charge.* This apparent increase in the charge would cause the broken ends of the larger less dense cross-sectional vortex to reconnect the thinner but denser ends of the smaller cross-sectional vortex.

Figure 6.7

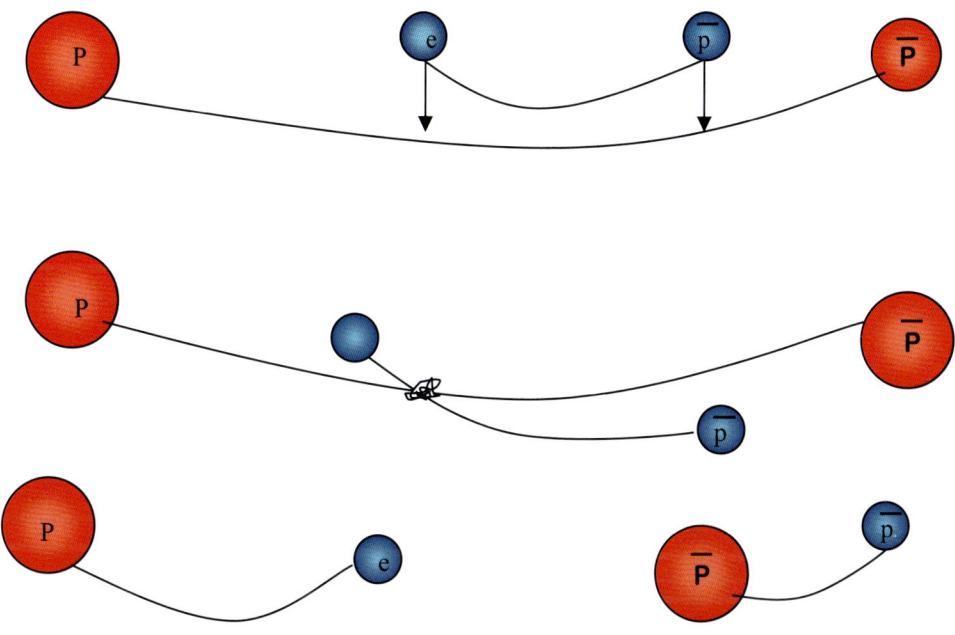

[Where P = proton, $\overline{P}$ = anti-proton, e = electron, $\overline{p}$ = positron]

Note: such actions may be responsible for some of the seemingly unexplainable motions of "particles" observed in the subatomic world.

Chapter 7
The Neutron

In the subatomic world of microscopic space, where the abnormal is normal and the strange is the order of the day, the neutron is the most bizarre inhabitant of all: the neutron is a hole within a hole!

It is easy to see how electrons and protons are three dimensional holes. This discovery is revealed because their electrical charges are created by three dimensional space flowing into or out of them. But the neutron has no charge. This means no space is flowing into it or out of it. So how can it be a hole?

The answer is that the neutron is not one solitary hole; instead, it is a hole within a hole! A simply fantastic concept!

The neutron is created when an electron is shoved up against a proton and completely encircles it; or, a proton is hit by an anti-neutrino, its higher dimensional vortex breaks and completely encircles the three dimensional surface.

Because the electron completely encircles the proton, the space flowing out of the electron is no longer flowing outward into the three dimensional space of our universe. Instead, its direction is reversed. It is now flowing inwards, directly toward the three dimensional hole, (the proton) the electron is surrounding. This situation creates an enclosed loop.

The space flows out of the proton and into higher dimensional space; as soon as it does, it fans outward into a cone shape, is turned inside out, and instantly curls back upon itself creating a tight loop. This tight loop completes the return back into three dimensional space by flowing directly onto the surface of the encircling electron, forming a fourth dimensional torus - or donut. A fantastic shape!

Also, just like the proton and the electron, the neutron is rotating around a fourth dimensional axis of rotation [see PART II #4]. This axis of rotation is responsible for its ½ spin, and its magnetic moment. The neutron is also surrounded by a sphere of less dense 3d space.

Just like the proton, the neutron also possesses a region of less dense space surrounding it. The proton within originally generated the less dense region that surrounds the neutron.

The region of dense space surrounding an electron does not exist around this electron because the space flowing out of the electron is no longer flowing towards three dimensional space. Instead, it is flowing in the direction of fourth dimensional space and is no longer pushing a volume of three dimensional space back into the three dimensional universe.

Figure 7.1

Just like all of the previous drawings, the sphere of less dense space surrounding the neutron is massive in comparison to the physical size of the neutron. Hence, this illustration is only a representation.

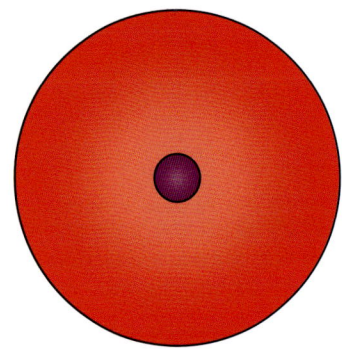

Because this extraordinary creation of nature we call the Neutron is so unique, an attempt to illustrate it was made in the following drawings. Unfortunately, since it is impossible to draw fourth dimensional space, these two dimensional to three dimensional sketches are used:

Figure 7.2

INITIAL CONDITION: anti-neutrino strikes a proton:

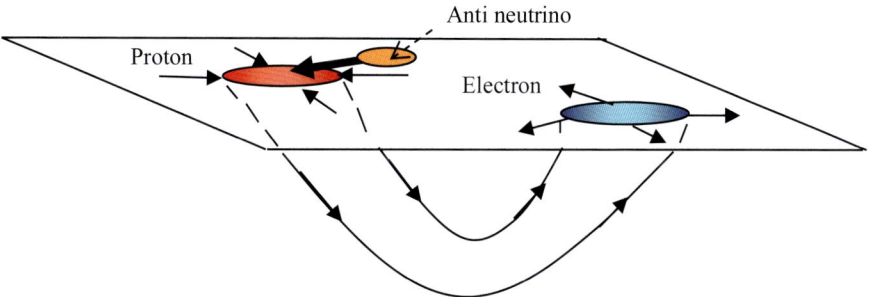

Figure 7.3 STEP #1: the vortex breaks: [Note, this break is much closer to the surface than seen here.]

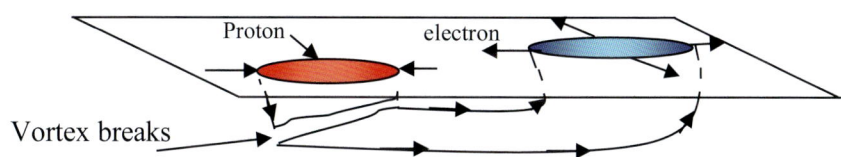

Figure 7.4 STEP #2: Isolating the proton from the above drawing, and expanding its size, note how the bottom of the vortex begins to curl outward:

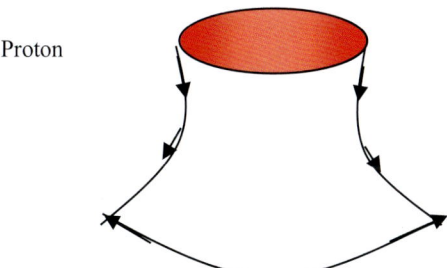

Figure 7.5 STEP #3: The curl becomes more pronounced as it continues to move upward:

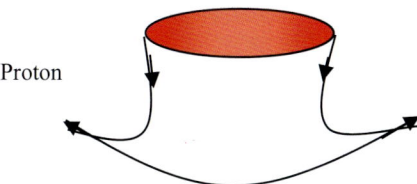

Figure 7.6 STEP #4: The vortex curls upward at an incredible speed (speed of light) towards the top of the hole we call the proton:

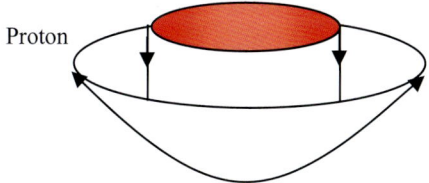

Figure 7.7 STEP #5: The vortex approaches the top of the hole called the proton:

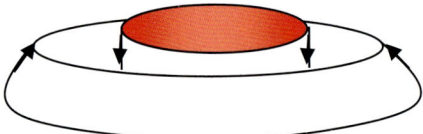

Figure 7.8 STEP #6: The vortex curls back into the hole called the Proton, forms a torus, the circulating flow begins and becomes a new "particle" science calls the Neutron.

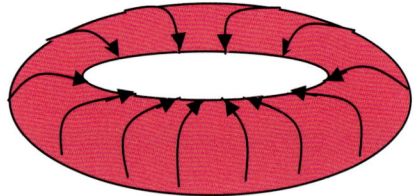

Figure 7.9 STEP #7: Another new "particle" called a *positron* is created when the end of the vortex attached to the electron reaches the two dimensional surface.

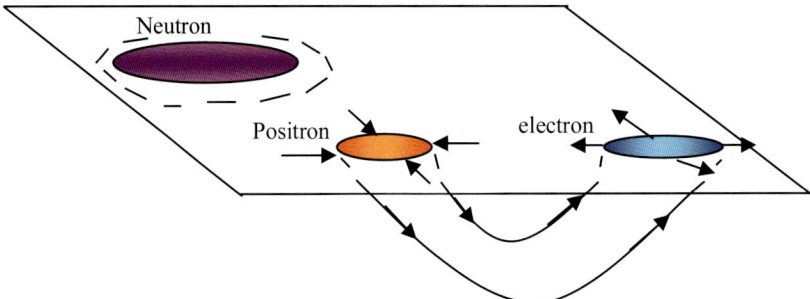

Note how space flows into the hole called the positron; turning it into a "particle" with a charge opposite to that of the electron. The neutron has no charge because none of the space surrounding the neutron flows into it, or out of it.

Also, the higher dimensional vortex is bent into a very tight loop. In this tight loop, the vortex is turned inside out, creating the weak force of nature. Hence, the neutron's "neutral" charge, and the weak force of nature are both explained, clearing up two more of the great mysteries of nature!

> The knowledge of the vortex and the realization that "particles" are really three dimensional holes at the ends of the vortex allows us to explain the mystery of science called the CONSERVATION OF CHARGE: see PART II #10

Chapter 8
Particle Collisions

In the subatomic world, "particle collisions" are nothing like the collisions of objects we are used to seeing. Nor are they like anything science presently imagines!

The explanation of the neutron now allows a closer inspection of the vortex using particle collisions.

Although the discussion of all of the particle collisions that are possible in the universe would take a volume of books to explain, the following are used only to show that the vortex is real and not just a hypothetical concept.

The discovery of the vortex is based upon the results of a collision between a proton and a particle called an anti-neutrino [explained in PART II #17], and the collision between a particle called a neutrino and a neutron.

In the world of Nuclear Physics, as we have just seen, when a proton and an anti-neutrino collide, the result is a neutron and a positron. This collision is viewed by the science of the 20th Century as a collision between particles – instead of holes, and is seen and taught in the following way:

Figure 8.1 The anti-neutrino approaches the proton on a collision course

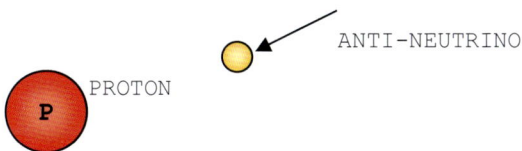

Figure 8.2 The anti-neutrino collides with the proton.

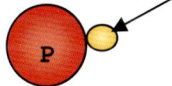

Figure 8.3 Two completely different particles emerge

In Figure 8.3, two completely new particles emerge from the collision! Instead of a proton and an anti-neutrino, we now have a particle called a neutron (N), and another called a positron! This result is not only confusing it is bizarre!

How can this be? There is no visual explanation [until now] for this phenomenon. There is a lot of talk about energies, Newton's laws, and mathematics. But as far as explaining why some "particles" turn into other "particles", no popular explanation is adequate.

Many years ago, when I was at college, such collisions were explained in terms of billiard balls. But the end results of these collisions in no way, shape, or form resemble the collisions of any type of billiard balls known. These explanations were never adequate or appropriate.

But the past is past. Using the new principles of the "Vortex Theory" this collision is easily explained:

Figure 8.4 The anti-neutrino approaches the proton

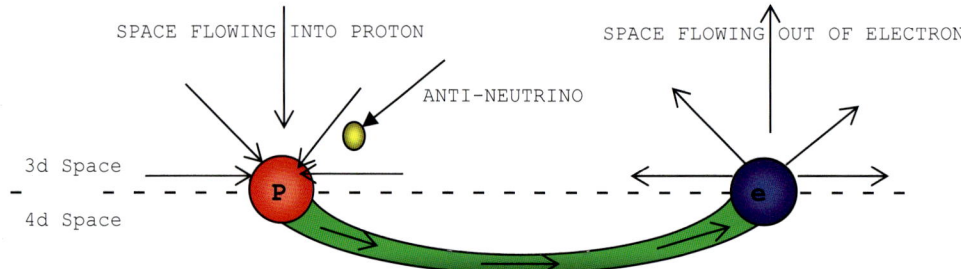

Figure 8.4, both the proton and its higher dimensional vortex are seen. Although we cannot see into higher dimensional space, this liberty was taken to illustrate the vortex and how it is connected to an electron or anti-proton existing *somewhere else in the physical universe.* [Note, the electron or anti-proton could be close by, a few miles away, or millions of miles away.]

Note too, how the dotted line represents the demarcation between 3d space (three dimensional space) and 4d (fourth dimensional space). Also, the arrows represent space flowing into the proton, and out of the electron.

Figure 8.5 The anti-neutrino collides with the proton:

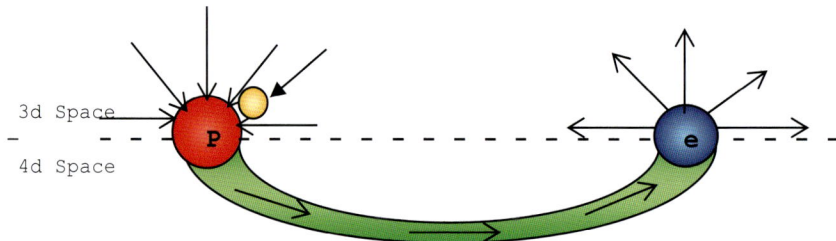

Figure 8.6

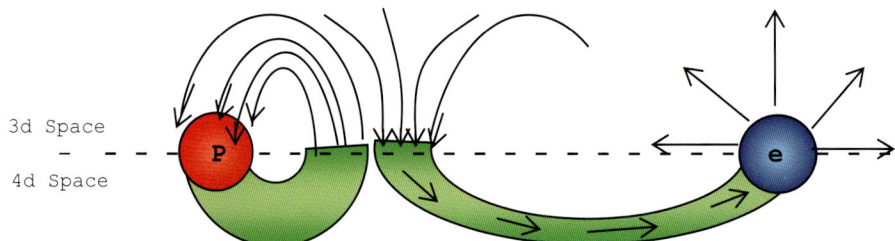

In Figure 8.6, note how the vortex breaks at the proton.

Figure 8.7

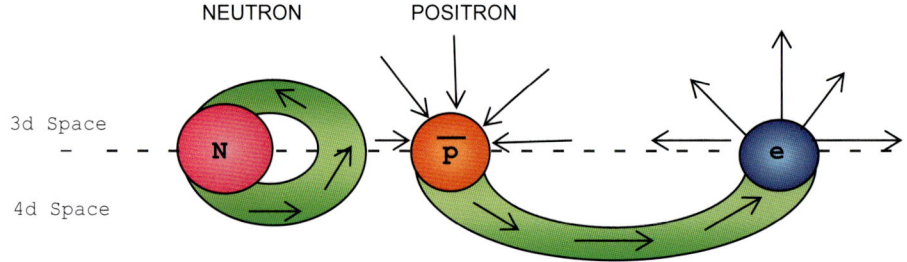

28

In Figure 8.7, when the vortex breaks, one end encircles the proton, creating the neutron, while the other end (which now has space flowing into it) becomes the positron.

Using this new way of viewing protons, electrons, and positrons it can easily be seen that when the proton is struck by an anti-neutrino, the higher dimensional vortex between the two particles broke, and both ends of the break terminated their connections within three dimensional space.

The end still connected to the electron became the positron, while the other encircled the proton. Although in the past, the positron was identified as the anti-particle of the electron, now, it can easily be seen that it is really just the other end of the vortex. Because space is flowing into it, it is also easily understood why it has a charge opposite to that of the electron.

However, the other "open end" of the vortex, [an "O*pen End*" is actually an electron], still connected to the proton performs a strange movement. It completely wraps itself completely around the proton, enveloping it. The neutron can now be seen for what it really is; one spherical hole completely surrounded by another spherical hole - a hole within a hole!

This hole within a hole forms a fourth dimensional donut, or torus. Although, as said before, a fourth dimensional torus is impossible to draw, an example of a three dimensional torus is a smoke ring.

The next collision diagramed below between a neutron and a neutrino, is most fascinating: to show that the neutron is indeed constructed out of an electron and a proton, the following collision between a neutron and a neutrino is analyzed.

As before, the contemporary view sees the following sequence:

Figure 8.8 The neutrino approaches the neutron on a collision course.

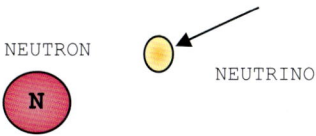

Figure 8.9 The neutrino collides with the neutron.

Figure 8.10

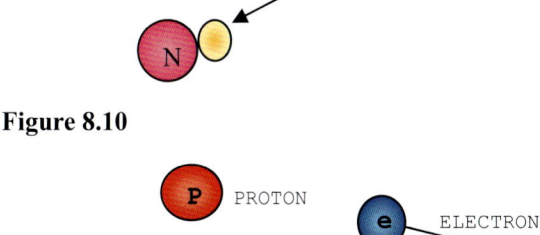

In Figure 8.10, when the neutrino collides with the neutron, "presto, whamo!" the collision produces a proton and an electron. And again, just as before, 20th Century science can only say that these results occur but cannot state why these results occur. However, when we use the Vortex Theory to analyze this collision, we see an entirely different set of circumstances:

Figure 8.11 The neutrino races towards the neutron.

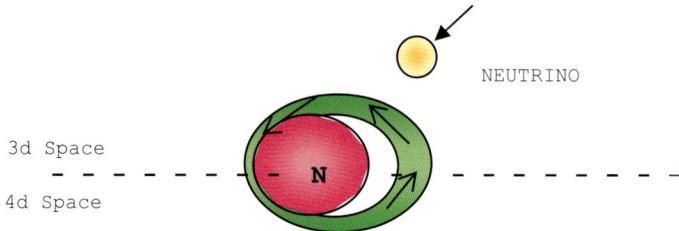

Although it is impossible to accurately draw, note how the higher dimensional vortex completely encircles the neutron.

Figure 8.12 The neutrino collides with the neutron.

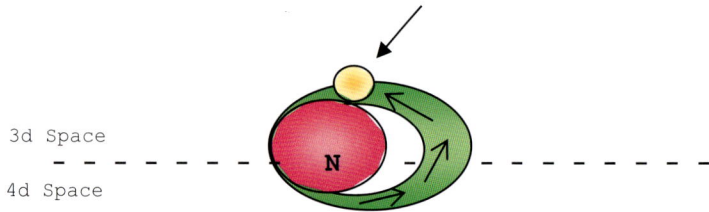

Figure 8.13

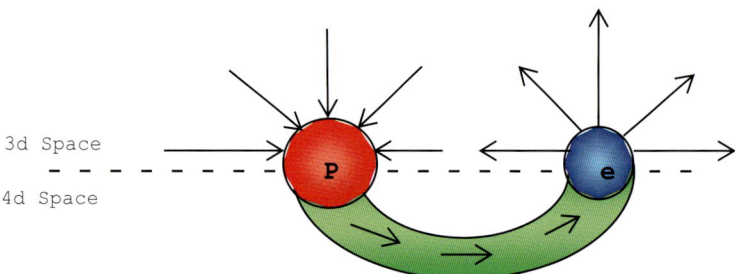

In Figure 8.13, the twisted vortex breaks free, liberating the proton inside, while the other broken end becomes the electron. Hence it is now easily *seen* that when the neutrino collides with the neutron, the end of the hole surrounding the proton collapses back into an electron – the three dimensional hole at the other end of the vortex.

The Conservation of Charge is easily *seen*. Because neither the neutron nor the neutrino had any charge, the net charge before the collision was zero. After the collision, because the charge on the proton is positive (+1) and the charge on the electron is negative (– 1), the net charge is still zero! Hence, because the charge before the collision and after the collision is zero, the Conservation of Charge is confirmed without the use of mathematical formulas.

This knowledge of the internal structure of the hydrogen atom has an added bonus - incredibly, it finally allows us to understand how a photon of light is constructed!

Chapter 9
The True Vision of Energy

> *Photons of energy are condensed packets of three dimensional space thrown from vortices. This condensed packet of 3d space both displaces the space surrounding it outward while it expands and contracts as it moves through the universe. The condensed region of space that is the photon creates the particle effect; the rotation or spin of photons creates their electromagnetic effect; their expansion and contraction creates their electric effect; while the region of denser space that is pushed outward and surrounds it as the photon passes through the universe, combined with the photon's expansions and contractions in the surrounding space, creates its wave effect!*

We all know that everywhere we look in nature we see an incredible variety of colors and shapes. However, what most of us don't know is that we never see anything! We don't even see the words on this page!

The only things we ever "see" are "photons" of energy.

Photons of energy are emitted from an energy source (such as the Sun, or a light bulb), hit the page, "bounce off", and hit our eyes. Without these photons we wouldn't be able to see anything at all. But just what are we "seeing"? What are these "photons" of energy? (What is energy?)

One of the greatest mysteries in all of physics is energy: what is it?

According to present day science, energy is contained within tiny particles called photons. These photons possess both particle and wave characteristics. But where they come from, what they are made of, or how they are constructed is unknown.

But not anymore!

With the discovery of the two vortices flowing back and forth between protons and electrons, the great mystery of the photon is finally unraveled - and a fascinating vision unfolds before us.

What contemporary science calls a photon is really a condensed portion of three dimensional space that is thrown out of the vortex. It is too bad that 19th Century believers in the Aether theory got it backwards; they believed that matter was a "condensation" of space (like ice floating in water). If instead, they had said that energy was a "condensation" of space, they would have been very close to the truth!

Although there are several ways to create photons, in the hydrogen atom, a photon is part of the volume of three dimensional space flowing from the electron to the proton. When certain conditions occur, the vortex shortens, throwing a packet of condensed three dimensional space back into the three dimensional space from whence it came.

The photon's velocity is also very important. The velocity of the photon is the velocity of the vortex, allowing us to deduce that the speed of light is a function of the speed of the vortex!

Although the present vision of the universe states that the photon "instantaneously" travels at the speed of light, we can now see that this is not true at all. It can readily be seen that a photon is but a part of the volume of three dimensional space in the vortex that was already circulating around and around at the speed of light before it was thrown free of the atom.

When a photon approaches an atom, is absorbed through the proton, and then is emitted through the electron. These principles can be seen in the following drawings:

Figure 9.1

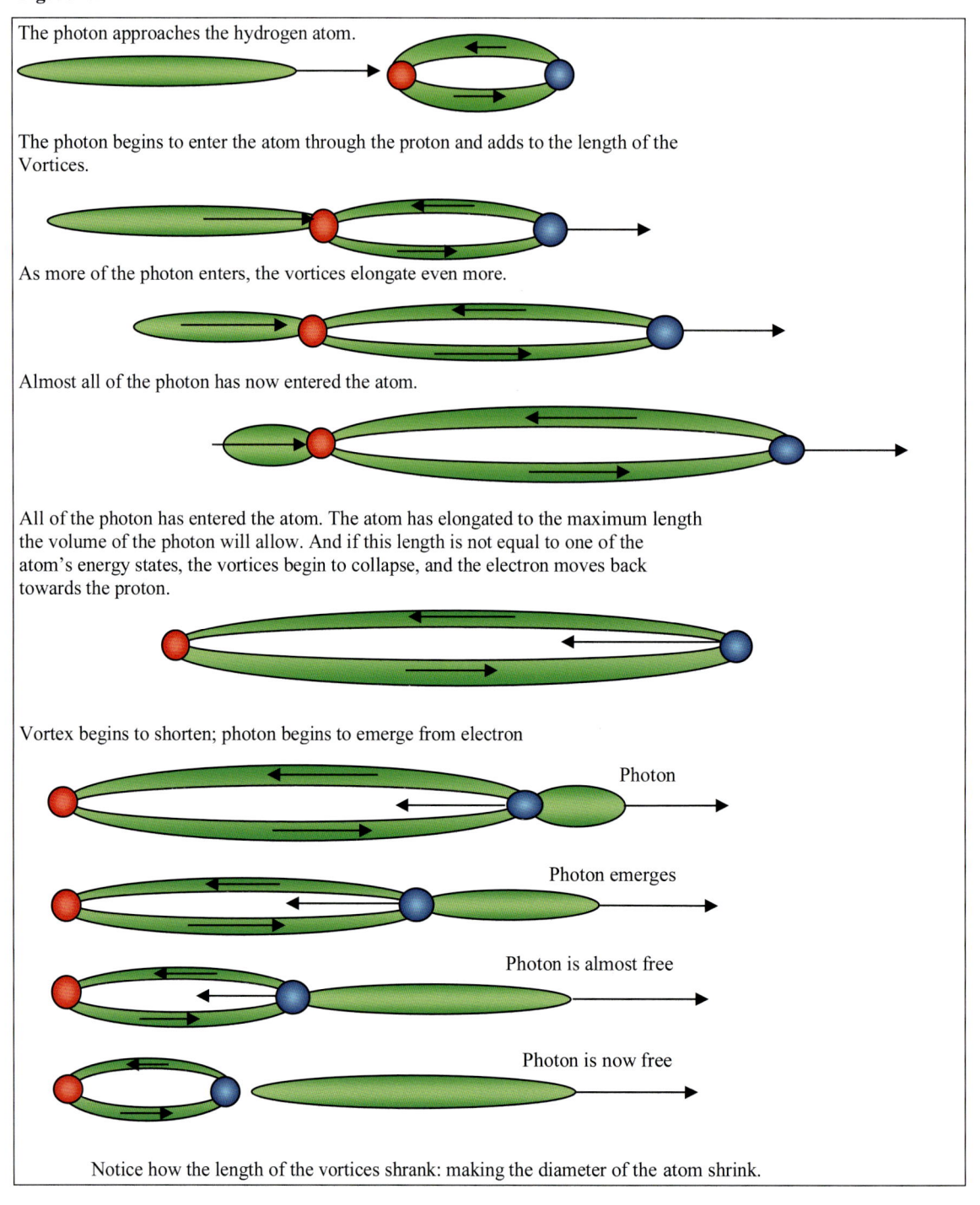

The photon approaches the hydrogen atom.

The photon begins to enter the atom through the proton and adds to the length of the Vortices.

As more of the photon enters, the vortices elongate even more.

Almost all of the photon has now entered the atom.

All of the photon has entered the atom. The atom has elongated to the maximum length the volume of the photon will allow. And if this length is not equal to one of the atom's energy states, the vortices begin to collapse, and the electron moves back towards the proton.

Vortex begins to shorten; photon begins to emerge from electron

Photon

Photon emerges

Photon is almost free

Photon is now free

Notice how the length of the vortices shrank: making the diameter of the atom shrink.

These illustrations not only allow us to see how a photon is absorbed and then emitted from an atom, but also, a little consideration allows one to explain many more of the previous heretofore mysteries of light reflection, refraction and absorption.

One of the most elegant relationships in all of nature that is *not* revealed by these illustrations is the speed of the absorption and expulsion of the photon. Because both the photon and the vortex are moving at the speed of light, when the photon enters the flow of the vortex it does not interfere with the flow; and it exits without creating any interference to the flow. This relationship never changes, even the speed of the atom is irrelevant. Such a harmony is not only elegant it is also very beautiful.

Great beauty is also found in the wave aspects of the photon.

When a photon is emitted from an atom, it is an extremely dense packet of space. Consequently, it displaces the surrounding space, creating a dense region of space around it much like the space surrounding the electron:

Figure 9.2

The particle and wave aspects of a photon. Note, the sphere of dense space that surrounds the photon is massive. Because of space limitations on this page, this drawing is only a representation of the huge sphere

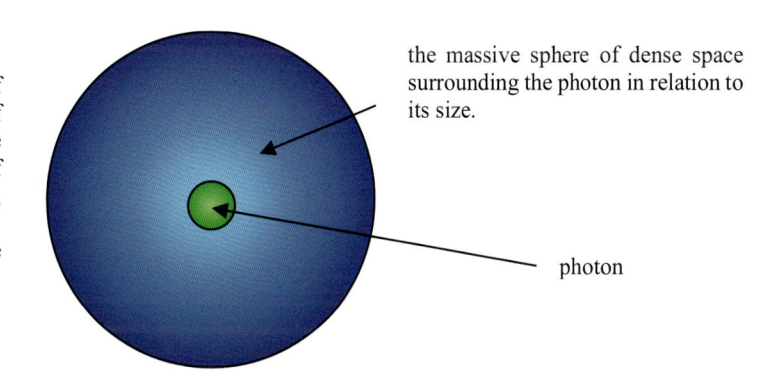

the massive sphere of dense space surrounding the photon in relation to its size.

photon

And again, just like the region of less dense space that surrounds the proton, the region of dense space that surrounds the photon is a function of the three dimensional space the photon is traveling through. This means that as the photon moves through the three dimensional space of our universe, the space the photon is passing through bends outwards away from the photon as it approaches, and then back inward as it passes by. Consequently, even though this massive sphere of dense space constantly surrounds the photon and seems to travel with it, *it is the region of space this dense region is passing through that is growing denser as the photon approaches, then less dense as the photon passes by.*

In 1865, James Clerk Maxwell proposed that light was an electromagnetic wave consisting of an oscillating electric and a magnetic field. And that as the electric field expanded, the magnetic field contracted and vice versa. It must be said that this was a brilliant idea, but Maxwell had no idea how these two fields were being generated. But that is all ended. Using the principles of the Vortex Theory, we can now give a visual presentation of how both of these fields are created.

The creation of the photon's *Electromagnetic characteristic* is a result of its spin. Because the electron is spinning [see PART II #4], when the photon is thrown from the vortex it is spinning too.

Figure 9.3 below represent two photons seen edge on and moving from left to right across the page. The first photon A, is spinning clockwise; while photon B is spinning counterclockwise. They are both traveling at velocity v, the speed of the vortex from which they were thrown. We recognize this velocity as "C", the speed of light. But it must be remembered that C is in reality merely a function of the velocity of the vortices speeding between protons and electrons.

Figure 9.3

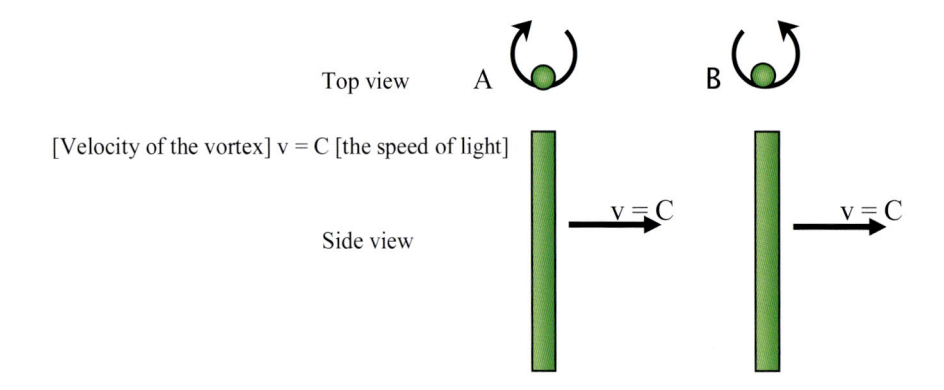

Top view   A ⟳   B ⟲

[Velocity of the vortex] v = C [the speed of light]

Side view   →v = C   →v = C

Because both photons are expanding and contracting perpendicular to the direction of travel, it is interesting to note that they can only have one of two spin orientations: clockwise or counterclockwise!

The Perpendicular Expansions and Contractions of the Photon are a most ingenious characteristic of nature.

The photon has to expand and contract outward in long tubular shapes rather than flattened out pancake shapes. If it expanded and contracted in a flattened out "pancake" shape, the velocity of space in front of it would have to exceed the speed of light to get around it, creating a "Rip" or "Tear" in the surface of three dimensional space.

Even though the speed of the expansions and contractions of the photon slightly exceeds the speed of light, the region of denser space that surrounds it allows this effect. Denser space has a higher elastic modulus that allows it to bend and flex faster.

The reason why the photon even expands and contracts is due to its attempt to blend back into the three dimensional space from which it came. As the photon races away from the electron it was emitted from, it tries to expand back into the three dimensional space it originally came from but instead finds that it can only expand in a direction perpendicular to its velocity of travel.

The expansion of this dense region takes place in a tubular shape allowing the space in front of it to swiftly move around it. As it expands upward and downward simultaneously, the space immediately above and below it is pressed upward and downward respectively. It is the pressure of this surrounding space that keeps the photon from expanding back into the three dimensional space it is made of. This event occurs when the photon, having expanded as far as its elasticity allows it, finds that the space immediately above and below it is now pushing back down upon it, causing it to contract back upon itself.

This contraction continues until the photon is forced back into the condensed spherical shape, beginning its expansion all over again.

As a photon travels through three dimensional space, its expansion and contraction perpendicular to the direction of its velocity creates its frequency. The length of the beginning and ending of these expansions and contractions creates its wavelength (in Figures 9.4).

Figure 9.4

Photon moving from left to right across page

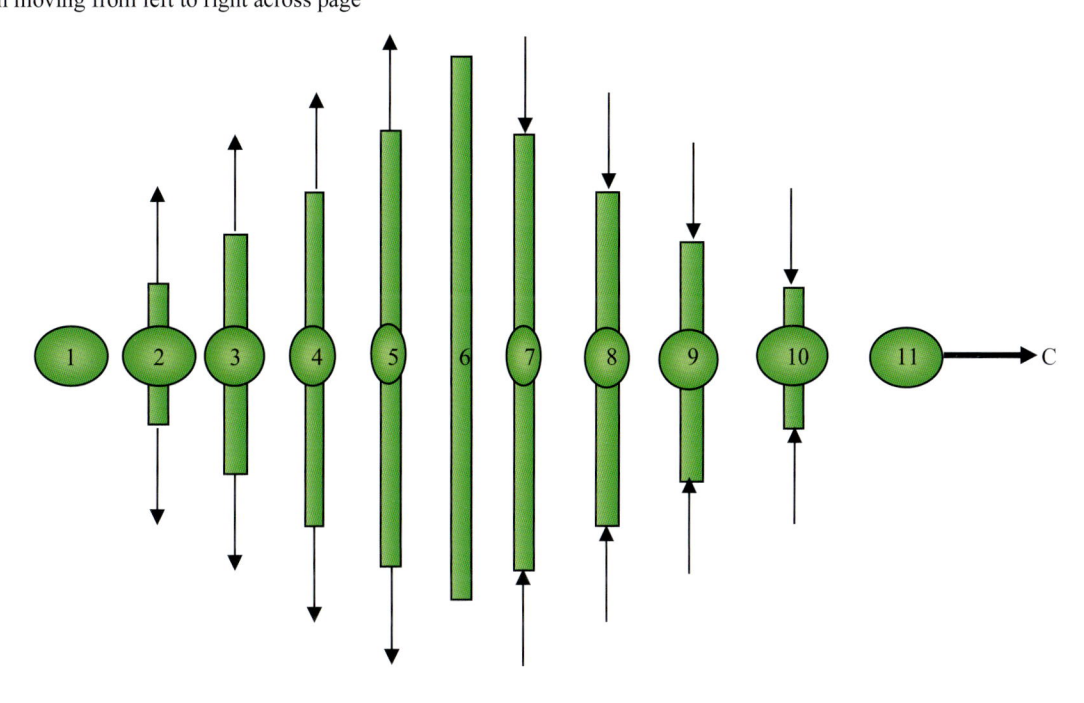

The volume of the denser space within the photon determines its length of expansion and hence its frequency.

The reason for oscillating exchange of the electric and magnetic fields is a result of the tradeoff between the changing diameter of the photon. When the photon is in shape #1, it is rotating and its radius is creating maximum rotation of the surrounding space. As the photon expands, it can be seen that this radius grows smaller and smaller, until at #6, the radius is very tiny creating only a very small amount of rotation in surrounding space. However, as it then contracts and goes from #7 to #11, it can be seen that its radius again increases, allowing it to again create a maximum rotation in surrounding space.

But at the same time, it can be seen that as the radius grows smaller, the height of the expansion increases. This expansion creates the electric effect of the photon. At #1, its height is at minimum while the magnetic effect is at maximum. While at #6, the magnetic effect is at minimum while the electric effect is at its maximum: visually allowing us to now physically observe Maxwell's brilliant deduction.

Note too: it should also be mentioned that the gamma ray possesses so much condensed space that when <u>*seen head on*</u>, it expands and contracts in a six sided "star pattern":

35

Figure 9.4

Figure 9.5

Seen sideways, it looks like this:

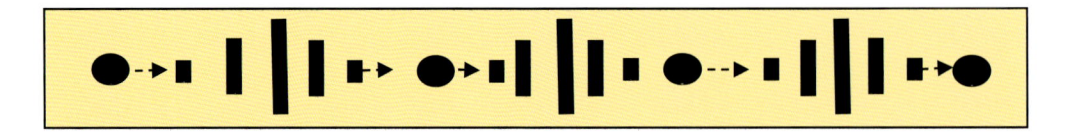

The discovery of how the photon is created and expands and contracts finally allows us to explain one of the most perplexing and enduring mysteries in all of science - the particle and wave characteristics of light: see PART II #1

Chapter 10
Creating the Forces of Nature

The forces of nature are created out of configurations of less dense, dense flowing space and moving space. As such, the forces of nature are manifestations of space itself. There is no unseen "substance" that somehow permeates the universe. Shockingly, the forces of nature do NOT attract matter to matter! Instead, distortions upon the surfaces of the "particles" matter is made out of accelerate them towards each other!

Nor is force carried by "particles." The shocking truth about force is that it just doesn't exist. It is not a mysterious substance surrounding matter binding it together. These are mistaken reasoning errors of past generations.

THE PREOCCUPATION WITH THE WORD "FORCE"...

Ever since Isaac Newton got hit on the head by an apple falling from a tree and discovered gravity, the world has had to deal with the presence of a mysterious unseen "force". This force, called Gravity causes matter to accelerate towards other matter. Even when we are standing still upon the seemingly motionless surface of this planet, the molecules within our bodies are constantly being accelerated downwards towards it. We call this acceleration our weight. It is this acceleration that holds us upon the surface of this spinning sphere and keeps us from flying off into space.

In the distant past, this preoccupation with force was unheard of. But today it isn't. With the scientific education we have received in school, we now live with force every day. Force is with us every second of our worldly lives and with the discovery of three additional forces, the word "force" has become a casual part of our vocabulary. We use this word freely and easily, even though we don't realize that the creation of force is still a total and complete mystery.

Although many physicists believe these four forces are all manifestations of just one universal force, they can't prove it. The greatest of them all, Albert Einstein, spent many years of his life trying to unite the four forces into one universal force. But he, like everyone else who becomes obsessed with uniting the forces of nature tried, and failed miserably. Too bad, he never knew he was on a fool's quest! He never knew that the individual forces of nature are not individual pieces of one "universal force".

Although the Theory of Quantum Mechanics claims forces are passed back and forth between particles by other particles, the Graviton, a hypothetical particle, supposed to carry the force of Gravity has not been discovered. In the trillions of particle collisions in the world's linear accelerators, no graviton has ever been seen. This idea of trying to explain everything that exists in the universe by using "particles" is a function of the mistaken belief that space was made of nothing resulting in a convoluted way of reasoning called "Particle Logic".

Getting back to force, unfortunately, the real reason why nobody can explain what force is, or how it is being generated, comes from the fact that the universe contains no mysterious "substance" called force. Nor are there any subatomic particles that "carry force" with them – this is a misunderstanding caused by an erroneous view of how both space and matter are constructed as we shall soon see. Amazingly what we are about to see is the shocking revelation that matter is not attracted to matter by unseen forces or particles. Instead, what is revealed is that matter is accelerated towards other matter by distortions in the surfaces of the "particles" it is made of!!!

Chapter 11
The Force of Gravity

> *In the theory of Relativity, Albert Einstein believed that matter was made of something, and space was made of nothing. And yet, he also believed that gravity was created by bent space surrounding stars and planets.* **But how can something made of nothing be bent?**
>
> *Fortunately, the truth is now known. Although space appears to be bent, this is an effect being created by a region of less dense space and is not the cause of gravity; it is not bent space that is creating the force of gravity. The force of gravity is a creation of less dense space.*

Most scientists and engineers are told in college courses that the force of gravity is created by a particle called the graviton. Nothing could be further from the truth!

The less dense regions of space that surround the holes ("particles") of matter create the force of Gravity. Einstein's hypothesis that bent space is equivalent to gravity is wrong. *It is less dense space that creates Gravity.* The "bent space" surrounding stars is an *effect* created by the addition of all the spherical shells of less dense space surrounding the protons and neutrons that account for the majority of the mass in stars and other large astronomical bodies.

When we stand upon this planet, this less dense region of space surrounding the Earth distorts the shape of every proton, electron, and neutron within our bodies towards the Earth's center. The collective attempt of these particles to straighten out - <u>pushes </u>us towards the surface of the Earth and becomes the "force" we identify as our weight. Note**: *we are not attracted towards the earth*;** the particles that atoms are made out of **_PUSH_** us towards the center of mass of the earth!

Figure 11.1

The Earth's Gravitational field: Although the red sphere represents the Earth's gravitational field, this is only a representation. Because the field is massive in relation to the size of the Earth, this smaller field was drawn to Earth fit the page.

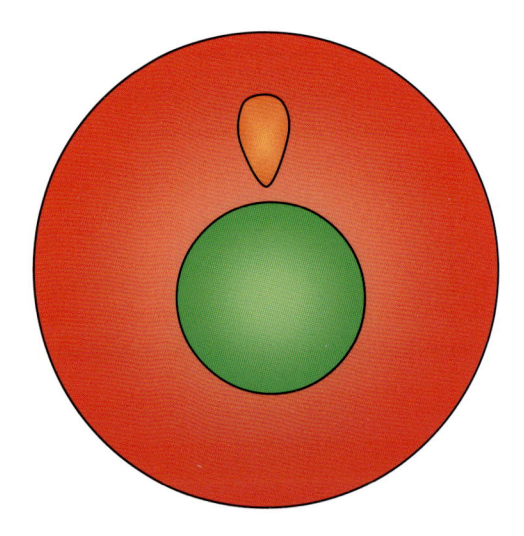

Note the distorted spherical yellow hole above the Earth. [out of proportion], this hole represents
a single proton. Although the proton is a three dimensional sphere, it is now distorted into a pear shape; causing it to be accelerated towards the earth.

Because the red sphere surrounding the Earth represents the three dimensional volume of less dense space that is stretched inwards towards the center of the Earth, the proton is now caught in this less dense region too. But there is now a problem.

As the three dimensional space surrounding the Earth is stretched towards the Earth, it cannot stretch across the fourth dimensional void within the center of the proton. Consequently, the side of the proton facing the Earth is stretched towards the Earth while the side facing away from the Earth still retains its spherical shape:

Figure 11.2

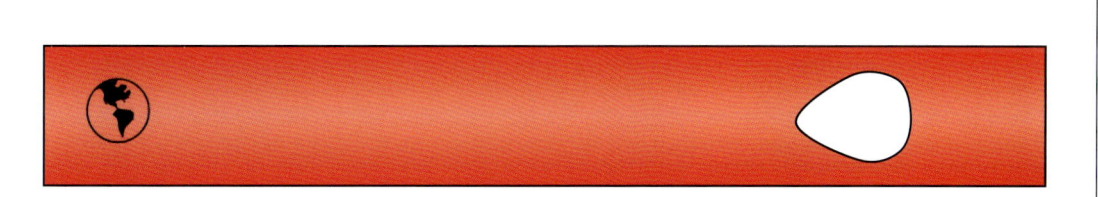

Although the above picture is a poor drawing, you get the idea. You can see how the shape of the proton is distorted in the direction of the Earth. This distortion of the three dimensional surface of the three dimensional hole is responsible for creating the force of gravity. It is now understood that gravity is not a "Pull"; instead, it can now be seen that surface distortions of the "particles" that make up matter are responsible for the movement of matter towards matter. As can be seen in the diagram below, the distortion of protons in the nucleus of two different atoms are "attracted" to each other by their mutual distortions:

Figure 11.3

(Note: these two holes represent two protons in the nuclei of two different atoms.)

#1   Normal shape   Normal shape

Length A

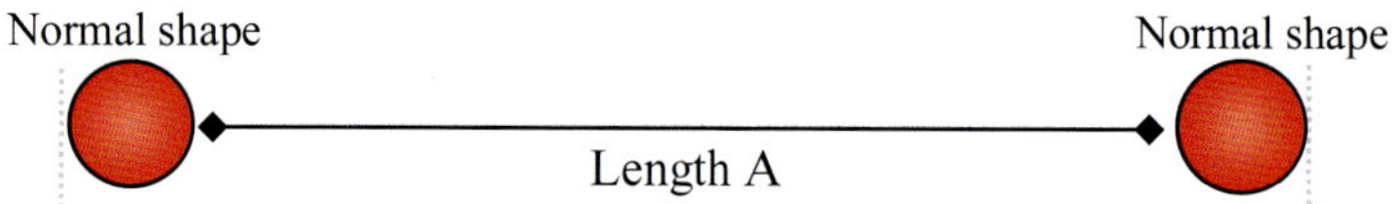

When two protons are near enough to each other, the less dense regions of space that surround them (these regions are not shown here) overlap, creating an *even less dense region of space directly between them.* This less dense region of space surrounding each sphere stretches the surfaces of the two spheres that are directly opposite each other, but cannot stretch upon the opposite sides of the spheres. This effect distorts the sides of the two spheres that face each other into "pear shapes":

#2

At the very tip or point of the pear shape, space is sharply bent. This stress creates a strain upon the front surface of the hole, forcing the back of the hole to move forward: causing it to move forward.

#3

As the back of the distorted hole moves forward, the spherical shape of the hole is again created: and the proton is now moving at velocity v_1.

#4

However, the instant the hole begins to move forward, regaining its shape, the hole is distorted again; surface stresses at a and b cause such severe distortion that the surface of the hole at point c moves forward again to reduce the strain.

#5

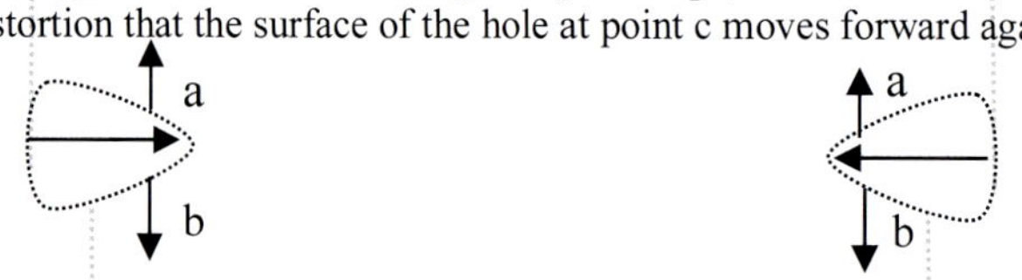

The moving hole returns to the shape of a sphere, and the whole process begins again. But this time, this additional velocity added to the initial velocity now creates an acceleration: $v_1 + v_2$ = acceleration. And as it continues to accelerate, the velocity gets larger and larger: $\sum = v_1 + v_2 + v_3 + v_4 + n$. Where n = number of distortions.

#6

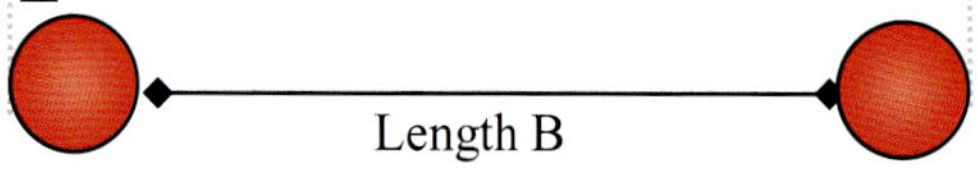

Length B

Length B is now shorter than length A. Equally important, because the distortion of the three dimensional hole distorts the fourth dimensional space within, and subsequently distorts the surfaces of the quarks, the same accelerations are created upon the quarks too. This subject will be discussed in Book 3, *The Quark Theory*.

As the process of distortion and reshaping repeats itself, length A becomes shorter than length B moving the two holes towards each other. As the distance between them shortens, the density of the space between them becomes even less dense. The decrease in density allows the hole in (Figure 11.3 #4) to elongate further, allowing the back to move upwards farther (Figure 11.3 #5) covering more distance, "accelerating" the holes towards each other.

The motions within the two protons disclose one of the great secrets of nature: the creation of internal waves within subatomic "holes"!

Because matter is made of holes instead of particles, it is extremely important to understand that a hole possesses an internal surface while a particle does not. *Hence, a hole can have a wave created upon its internal surface* while the particle cannot.

The realization that matter is made of holes reveals a wealth of information about the universe. These waves created UPON the three dimensional surfaces WITHIN these three dimensional holes are responsible for the creation of a number of previously unexplained phenomena. Perhaps the most noteworthy are NEWTON'S THREE LAWS OF MOTION, MOMENTUM, and THE CONSERVATION OF MOMENTUM [see PART II #6, #7, #8]

It should also be mentioned that the initial resistance to the creation of these waves is responsible for the creation of MASS and INERTIA [see PART II #5]. No Higgs Boson particle or Higgs field is needed. The explanations of all of these vital and most important mysteries of the universe are also found in PART II.

Note: Even though a region of denser space surrounds an electron, when it is caught in the gravitational field of a star or planet, its three dimensional hole will distort exactly like the proton. Consequently, it too will be accelerated towards the less dense space surrounding the star or planet.

Because quarks are fourth dimensional holes existing upon the surface of fifth-dimensional space, it is much more difficult for three dimensional space to distort their surfaces. However, because the fourth dimensional space within the three dimensional hole is itself slightly distorted by the distorted shape of the three dimensional hole, these fourth dimensional holes are slightly distorted too, causing them to also move in the direction of the distortion.

It should also be mentioned that if the proton is within an object that is for example resting upon the surface of the Earth, the proton will continue to be distorted and will continue to try to accelerate towards the center of mass of the Earth (<u>pressing</u> it against the surface of the Earth).

Finally, it should be noted that the distortion seen in Figure 11.3 also applies to a certain extent to the photon. Even though the photon is constructed out of a dense region of space, the distortion created by the less dense space surrounding the sun goes right through it. Nevertheless, because this region of less dense space becomes less as distance increases ($1/d^2$) even though the effect is very slight, this region of less dense space surrounding the Sun will be slightly greater on one side of the photon than the other. This will distort it ever so slightly in the direction of the Sun and make it move closer to the Sun as it flies past.

This distortion will create the famous effect of making the light seen from stars during a total eclipse of the Sun appear to move inward towards the Sun. It also created the "lens effect" seen in deep space above and below galaxies. Therefore, it is not the warpage of space-time that bends light around large heavenly bodies as Einstein proclaimed!

Chapter 12
The Electromagnetic Force

The electromagnetic force consists of two parts: the magnetic and electrostatic force. Both of these forces are created by flowing or rotating space: mistakenly called "lines of flux" by people who did not know what they were looking at. The magnetic force is created by the rotation of three dimensional space; the electrostatic force is created by space flowing into and out of charged particles.

THE ELECTROSTATIC FORCE...

The electrostatic force is created by space flowing into or out of a three dimensional hole. The electrostatic force is created by three dimensional space flowing into the proton and out of the electron.

Figure 12.1 Electrostatic Force

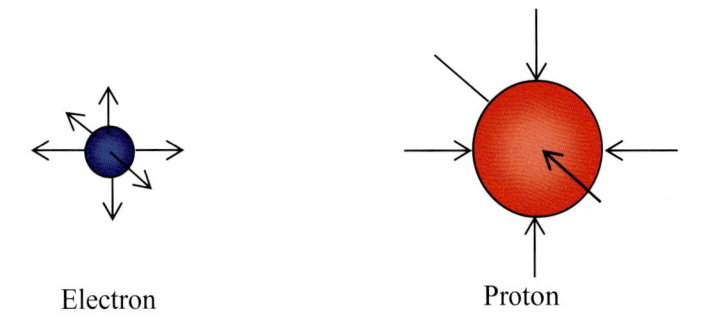

Electron   Proton

THE ELECTROSTATIC "ATTRACTIVE" FORCE...

The electrostatic "attractive" force develops when an electron and a proton come into a close enough proximity to each other that flowing space from one hole begins to flow directly into the other hole. The mutual motion towards each other occurs because the space flows more readily out of the electron in the direction of the proton – distorting the electron in the direction of the proton – causing it to move in the direction of the proton; while the flow into the proton occurs more easily in the direction of the electron – distorting the proton in the direction of the electron – causing it to move in the direction of the electron. It is the distortion of each "particle" that moves them towards each other, and NOT the attraction caused by some mysterious force!!!

Figure 12.2

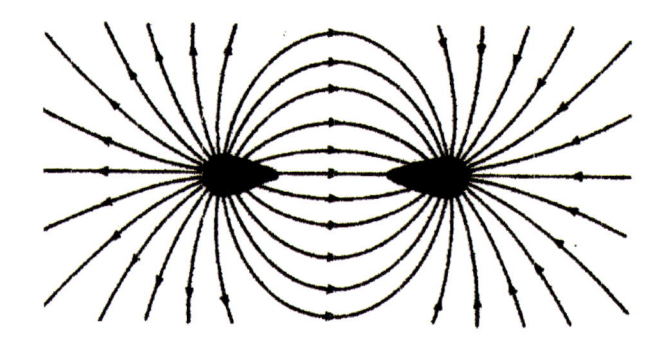

THE ELECTROSTATIC REPULSIVE FORCE BETWEEN TWO ELECTRONS...

The electrostatic repulsive force develops when two like holes such as two electrons come into close enough proximity to each other causing the flow of space directly between each hole to be "twisted" outward. For example...

When two electrons come into close proximity to each other the outward flow of space from one hole pushes against the space flowing out of the other hole. This resistance makes the region between each hole appear to be a denser region of space. This seemingly denser region causes the side of each hole directly opposite to each other to bend outward easier than the side facing it. This creates a pear shaped distortion in the surface of each hole. These distortions cause each hole to try to straighten back out into a sphere: in effect, causing each hole to accelerate in the opposite direction to each other; forcing them to move away from each other. Note: the greater the charge, the greater the distortion, and the greater the acceleration.

Figure 12.3

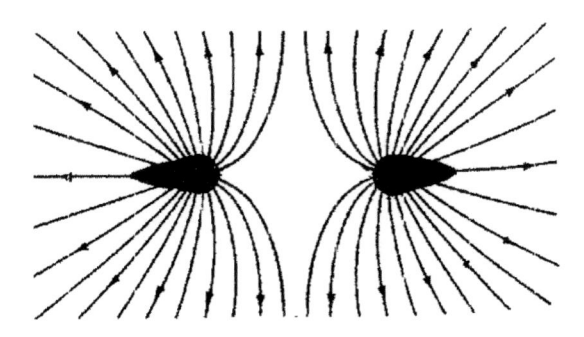

THE ELECTROSTATIC REPULSIVE FORCE BETWEEN TWO PROTONS...

Although it is easy to see why two electrons "push" themselves away from each other, why two protons repulse each other is not readily apparent.

When two protons come close to each other, it first appears as if the space flowing into each will pull them into each other. However, the reason why they don't come from the fact that the three dimensional space between them cannot flow in two opposite directions simultaneously. Consequently, to be able to flow into both particles along the x- axis, space has to now flow downward from the direction of the +y axis, and upward from the direction of the −y axis to get into each hole. This situation creates the exact same effect as when space was flowing out of both electrons creating a denser region of space between them. Because space does not flow into one side as easily as the other, the space directly between the two protons now creates the effect of a denser region of space. Hence, the opposite sides of each hole bend out easier than the sides facing each hole. They distort outward easier, and this outward bend accelerates them away from each other.

Figure 12.4

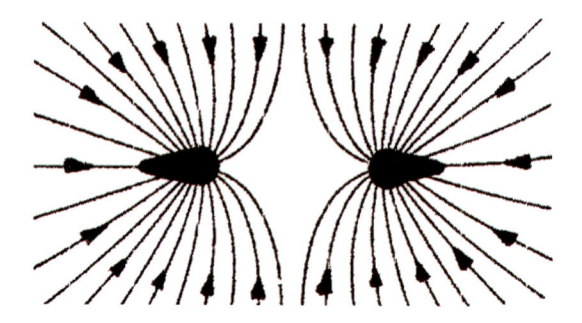

THE MAGNETIC FORCE…

Fortunately, for the magnetic force, there is nothing revolutionary to propose. It is fairly easy to understand, that Magnetic fields are created by rotations of three dimensional space around electrons. The spin of electrons creates rotations in space about them creating magnetic fields. It should be noted that the "attraction" and "repulsion" of magnets create the same distortions within protons and electrons as those seen in Figures 12.2 to 12.4 above. The additions of their spins in bar magnets create the famous lines of flux seen below in Figure 12.5.

Figure 12.5

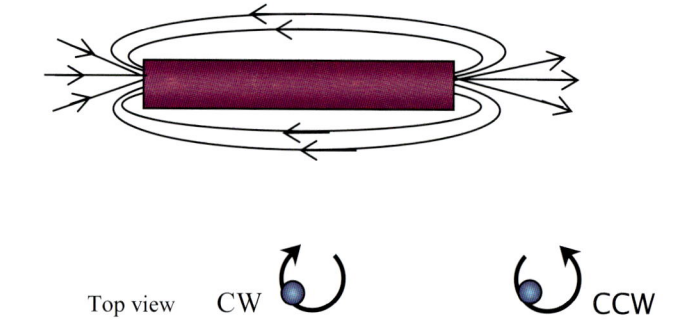

Figure 12.6

Top view   CW   CCW

The two spin states of electrons above are either clockwise or counterclockwise.

The electromagnetic field about a conductor…

When current flows in a wire, free electrons in the wire move from one atom to the next. As the electron moves, a point is reached where it breaks contact with one vortex while at the same instant begins to make contact with a vortex from the next atom along its line of travel.

During this brief instant, some of the space flowing outward from the electron is allowed to stretch out to other atoms lying transverse to its line of travel. It is the combination of all of these momentary flows that is responsible for the electromagnetic field about a conductor. The direction of the flows (left hand rule) is determined by the spin of the electrons. The spin of all the electrons is the same because their magnetic moments are aligned to the direction of the electric potential connected to the wire. The strength of the flow – the amount of flowing space – (magnetic field) depends upon the amount of electrons in transit at any particular instant: which is a function of the amount of current in the wire (Coulombs per second).

Chapter 13
The Weak Force

The weak force of nature is not really a force. The weak force of nature is associated with the neutron and how it "decays" into a proton, an electron, and an anti-neutrino. However, the "decay" of the neutron is not caused by a "force", but rather by the break-up of the three dimensional vortex surrounding the proton at the neutron's center! This break-up is caused by harmonics within the tightly bent vortex as it whirls in and out of the proton at the speed of light. These harmonics are created by the space bent inward around the proton opposing the space bent outward around the electron. These two different volumes are diametrically opposed to each other and seek to escape from each other.

THE TIGHTLY BENT LOOP OF THE VORTEX...

The weak "force" is a result of the higher dimensional vortex being inverted into a tight loop. Because this loop is in the fourth dimension it is impossible to draw, therefore the following is only a representation.

Figure 13.1

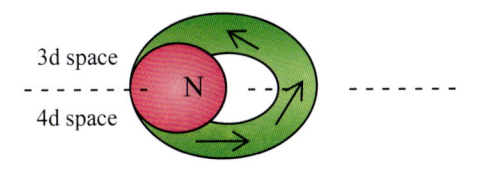

When this tight loop of flowing three dimensional space is broken, the ends of the vortex are again separated, allowing the proton and the electron to seemingly magically reappear.

What causes the vortex to break is denser space surrounding the electron trying to move outward and away from the space bent inward surrounding the proton. Because the tight loop is an unnatural bend, the space within the vortex is stretched more on the outside than it is on the inside of the loop. This stretched condition makes it less elastic, decreasing its elasticity.

Within the vortex, this decrease in elasticity makes its inside bend want to flow at a faster rate than its outside bend. This creates a stress between the inside edge of the flow and the outside edge. As the inside of the vortex tries to flow faster, it tries to pull away from the outside of the vortex, and eventually, the construction of the vortex is unable to handle the strain, and it breaks along its outside edge...

Figure 13.2 **Figure 13.3**

In Figure 13.2, note how the vortex breaks at the neutron; then in Figure 13.3, how the proton reemerges at one end of the vortex while the other end of the vortex becomes the electron, as an invisible electron-antineutrino flies off.

When the vortex breaks, the two ends of the vortex reappear upon the three dimensional surface as the proton and the electron. The anti-neutrino "particle" is created by the deflation of the volume of the three dimensional space that the larger neutron filled [explained in Book 3]. The whole process can be compared to twisting a strong spring into a sharp bend, letting go of it, and watching it snap back into its original shape.

Even though the weak force is more of a disturbance than a force, it is interesting to note that both the electromagnetic force and the weak force are related because they are both created out of flowing space.

Also, when the neutron is alone in free space, it only lasts about 10.3 minutes before it "decays" into a proton, an electron, and an anti-neutrino. But when the neutron is within the nucleus of an atom, it lasts much longer. The reason why it survives so much longer is explained in Book 3.

Chapter 14
The Strong Force

The strong force of nature is not only the strongest force of nature; it is perhaps the strangest force in nature. It is possibly the most unique of all the universe's marvelous sub-atomic, "Midnight World of Microscopic Space's" creations! It not only holds protons and neutrons together within the nuclei of atoms, it is responsible for creating the alpha particles thrown out of the nuclei of massive atoms.

THE CREATION OF THE STRONG FORCE…

According to the Vortex Theory, a neutron is a three dimensional hole surrounded by a spherical region of space bent into it; a hole within a hole.

Figure 14.1

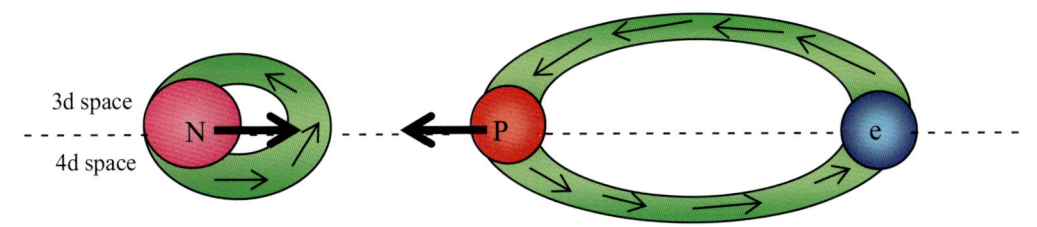

When a proton approaches a neutron, the less dense space surrounding each causes these two holes in space try to bend into each other. When they do, some of the space flowing out of the spherical hole surrounding the neutron tries to flow into the proton that is beside it, see the thick black arrows from neutron to proton in Figure 14.2.

This situation causes the enclosed vortex encircling the neutron to break free and encircle the proton in Figure 14.3 turning it into the neutron, while the neutron it vacated becomes a proton again. However, just as soon as the switch occurs, the process instantly begins all over again, causing the reverse to occur. The encircling vortex breaks free of the proton it now surrounds and returns to the one it just left - only to instantly repeat the process all over again: see Figure 14.4.

Figure 14.2

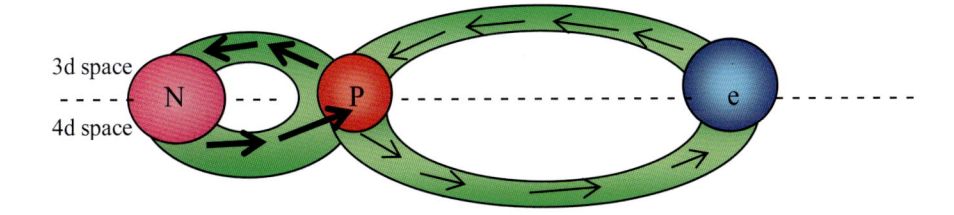

Figure 14.3

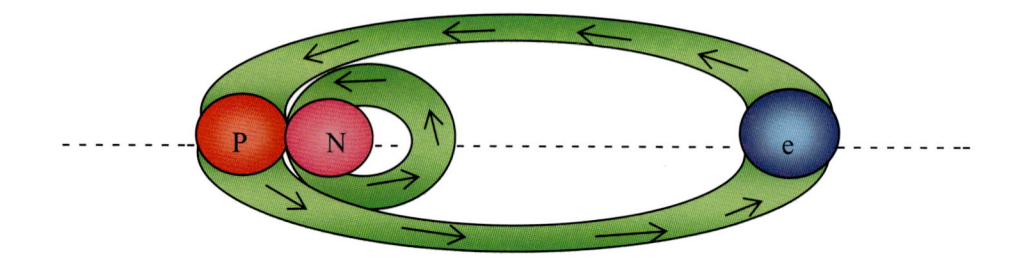

Figure 14.4

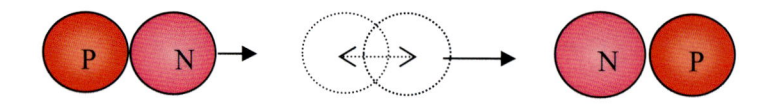

These two "particles" keep on doing this identity dance, back and forth. Because this reversal happens so incredibly fast, these dancing holes of space end up keeping one "particle" pressed tightly against the other - becoming the strong force of nature.

> In Book 3, it will be shown that an UP quark is passed to the old neutron and a DOWN quark is passed to the old down proton; reversing the quark content of each. As the two quarks pass each other, they create Hideki Yukawa's virtual particle.

[Note: when two protons collide, the attempt by three dimensional space to form only one surface appears to be the reason why two positively charged particles can temporarily bond together, forming the *delta particle* - a very short lived particle with a charge double that of a proton.]

It should also be mentioned that the neutron lasts longer within the nucleus of an atom than in free space due to the fact that it is constantly being recreated again and again. Because of this process, the stresses within its bent higher dimensional vortex do not have the opportunity to develop as fast as they do when the neutron is in free space.

48

Chapter 15
The Anti-gravity Force Using Buoyancy as an Example

*Contrary to present belief, there are not just four forces in nature but a fifth one too. Because it is contrary to the other four "forces", this fifth force can be designated the **anti-gravity force**. This anti-gravity force is responsible for a number of unexplained phenomena that occur in nature. Someday, when this force is artificially created it will not only be the stunning confirmation of the Vortex Theory, but also, it will be the greatest technological achievement ever made. It will allow us to finally travel to the stars, and will make atomic energy obsolete!*

Because nothing is presently known about this force, the phenomenon that it is responsible for creating will be revealed. Incredibly, one of the most famous and well known of the effects the anti-gravity force is responsible for is Buoyancy!

We all know about the phenomenon called Buoyancy, this is the reason why boats float in water; however, until now, no one knew why this phenomenon existed.

Buoyancy is important to us all. Without buoyancy, no ship could sail the sea, nor could anyone pan gold. Buoyancy is another one of those phenomena of nature that is so old, its acceptance is without question. We know that it exists, and because it is a phenomenon that is so common, we don't even think to question it.

But why does buoyancy exist? What is the mechanism in nature responsible for making an object "buoyant"? What makes something float upon the surface of the water? What makes minerals of different densities separate apart from one another in a revolving fluid such as those in a centrifuge?

The answer is something nobody has ever suspected before.

It is not the mass of the object (such as a piece of wood) that causes buoyancy, but the density of the space within the wood! It works like this: the *more* protons and neutrons per cubic inch, the *less dense* the space within; and the *fewer* protons and neutrons per cubic inch, the *denser* the space within. Therefore, it can be said that within a cubic inch of for instance a mineral containing a lesser number of protons and neutrons than another cubic inch of a different mineral containing more protons and neutrons, regions of denser or less dense space are in effect "trapped." This relationship can be seen in Figure 15.1, A and B: cross sections of rocks.

Figure 15.1

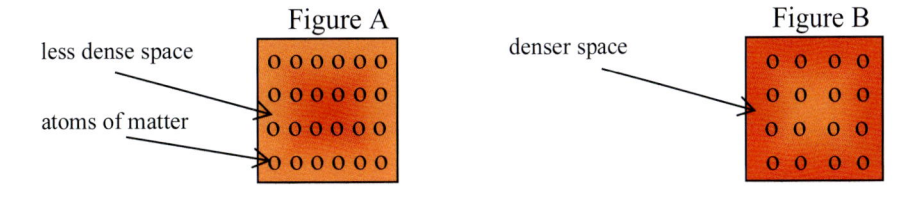

Figure A — less dense space, atoms of matter

Figure B — denser space

In Figure A, more matter per cubic inch creates a region of less dense space in comparison to Figure B. In Figure B, less matter per cubic inch creates a region of more dense space in comparison to Figure A. Within a stream of flowing water, when these different regions of space are being mixed together, or allowed to rotate together, the turbulence of the water allows them to rearrange their locations. The more dense regions of space found in less massive rocks try to move upward, while the less dense regions of space trapped in more massive rocks move downwards. In effect, *when moving*, these denser regions of space possess anti-gravity properties.

It must be emphasized that even though all of the protons and neutrons from both cubic regions of space are accelerated towards the Earth's center of mass equally, when mixed together, it is the denser region of space that is bent upwards and away from the direction of the Earth's center of mass. Consequently, this denser region of space that is trapped within a less dense rock seeks to move upwards and away from the Earth creating a buoyancy effect.

This same effect is produced within the hull of a ship. A region of more dense space is "trapped" within the empty hull of the ship making it push up out of the water.

A hot air balloon rises because *photons are dense regions of space*. When a massive amount of photons are trapped within an enclosure, such as a balloon, the region of air that is constantly exchanging them has in effect entrapped a volume of denser space. Because the surrounding space is less dense, the denser region of space inside of the balloon is accelerated upward.

THE UNIFORM DISPERSAL OF ONE GAS WITHIN ANTOHER...

Another example of anti-gravity effects upon the earth that no-one knows about is the uniform dispersal of one gas within another.

In solids, atoms and molecules are in fixed positions. Between these atoms and molecules are trapped regions of dense or less dense space. It is these dense and less dense regions that create buoyancy. But within gasses, a different situation is occurring.

Within gases, there are no trapped regions of dense or less dense space. All of the atoms within the gas are pressed against one another. This means the electrons in the outer shells of the atoms are all pressed together but not bonded to each other. Because of this fact, the bent outward regions of space surrounding the electrons in the outer shells of the atoms repel each other.

This repelling effect causes all of the atoms to move apart from one another, to disperse throughout the gas, and seek positions where the effect is at its minimum.

One of the most important questions that 20th Century science has never satisfactorily answered is how come atoms do not merge into other atoms? What keeps the nuclei of atoms from just piling up and creating one monstrous atom?

The answer is the denser space surrounding the electron. This region of denser space repels the denser space surrounding other electrons in other atoms and keeps them apart. In effect, Anti-Gravity keeps atoms from merging into each other. It also is responsible for the acceleration of galaxies away from each other as revealed by the Hubble Telescope

In the late 1990's, a picture taken from the Hubble telescope seemed to indicate that a supernova in a Galaxy located in a distant part of the universe was accelerating away from us at a much faster rate than it should. But how can this be? How can all of the previous spectrographic analysis of the pictures from the last 50 years showing the Red Shift of the Galaxies indicate one result, and this picture indicate another?

The answer to this dilemma can be resolved when it is realized that this picture was taken on the other side of a region in the universe where no galaxies can be seen with earth bound telescopes. This indicates that there is a vast sea of space between the Earth and this distant Galaxy.

In these vast reaches of space where there are few Galaxies, the space is denser. Hence, it tends to cause all of the matter in the Galaxies located on either side of it to be distorted in the direction opposite to it, causing these Galaxies to accelerate away from this vast wasteland towards other less dense regions of space. This phenomenon is responsible for the effect called Dark Energy explained in Part II, #14.

Chapter 16
Mass… [There Is No Higgs Boson Particle]

> Several years ago, we were told that CERN discovered the Higgs Boson Particle: the particle that explains mass. It was touted as a great discovery. Much celebration was done. There was the drinking of many bottles of Champaign. Higgs later received the Nobel Prize in physics. There is only one problem: there is no Higgs Boson particle! It does not exist! CERN got it wrong! Today you will learn the truth about the Higgs and how the phenomenon of mass is really created!

Supposedly, one of the greatest discoveries in the 21ST Century's science of physics was CERN's finding the Higgs Boson Particle: the particle that some scientists believe explains "mass" and completes what is called the "Standard Model", a so-called collection of "particles" used by particle physicists to explain the construction of the universe. However, nothing could be further from the truth: because how can you discover something that does not exist!

Up until 1964, science had no explanation for the phenomenon associated with matter called mass. However, in that year, a Physics Professor named Higgs proposed the existence of a new particle in nature called the Higgs Boson [that he graciously named for himself!]. Because nobody had any other explanation for mass, this seemed as good as any. Later on, clever European scientists decided to build a giant accelerator called the Large Hadron Collider at a complex called CERN on the France - Swiss border to search for this particle that would explain mass and other research.

Twenty two nations donated billions of dollars to the project and three thousand scientists went to work. It took a number of years to build CERN and it cost a lot of money [about 13.5 billion dollars]: money that some poorer nations such as Italy could not afford. They wanted out of the project. However, in 2011, when CERN needed more of its expensive funding or be shut down, Rolf-Dieter Heuer the then director of CERN suddenly announced that CERN was, "Close to discovering the Higgs particle!" See the below web site:

https://home.cern/science/physics/higgs-boson

> However, according to Dr., Prof., Konstantin Gridnev: then Chairman of the Nuclear Physics Dept. at St. Petersburg State University in Russia, (whom I had briefed on the explanation of mass and the fact that there was no Higgs particle), told me that he talked to a friend that worked at CERN after hearing the public statement stating: "That they were close to finding the Higgs" and was subsequently told, *"That they were not close to discovering anything!"*
>
> I was shocked and wanted CERN investigated but Konstantin was afraid they might be shut down. However, my dear friend Konstantin has passed on, and this false vision of mass is now being taught in universities and colleges all over the world. This has now gone too far and must be exposed as incorrect science; if not, millions of future scientists and engineers will now be taught false information, and it could take another 100 years before the truth is finally revealed!

Nevertheless, because of their reported, "close to a breakthrough", CERN received another billion dollar year of funding; and its 3000 employees another year of pay checks. And then again in 2012 when CERN finally faced the problem of either discovering the Higgs, or being permanently shut down by many of the poor countries in the treaty – [who signed on before finding out just how expensive it would become] - scientists at CERN suddenly, and miraculously discovered the Higgs Particle"! WOW!

However, how can you find something that does not exist? And if it does not exist, how can CERN's 2011 statement be true: because CERN's scientists could not be close to finding anything that doesn't exist!

Anyway in 2011, CERN's statement got them another year of funding: "a billion dollars" of funding! Another year of pay! What a great discovery this was! All of the permanent 3000 employees were overjoyed because they were not going to be fired. But the 22 countries funding CERN were not overjoyed, because they had to cough up another billion dollars!

Anyway, the next year in 2012 at CERN, when it was announced that the Higgs had been discovered, there was much celebration and the drinking of many bottles of Champagne.

But there is just one problem with this so-called great discovery, years before, when the principles of the Vortex Theory were applied to the phenomenon of mass; it was revealed that *no "particle" was responsible for the creation of mass*! Instead, it was discovered that it was the distortions in the three dimensional space surrounding "particles" of matter such as protons and electrons that are responsible for creating resistance to acceleration. That in fact, no "particle" was needed to explain mass.

WHAT ARE THE IMPLICATIONS IF THIS HIGGS BOSON MISTAKE IS NOT CORRECTED?

This is not just a game of "Catch the Crook!" The future of the next 100 years of science is at stake. For example:

Einstein's colossal mistaken belief that space is made of nothing, being subsequently taught in all of the colleges and universities in the world has forced millions of past and present day scientists and engineers to *assume* that everything that exists in the universe has to be made of particles: creating what can only be called a convoluted system of reasoning called "Particle Logic". This particle logic has lasted for over a hundred years! It has caused all of the scientists in the world to reach false conclusions about the data they have collected. And the exact same thing can continue for the next 100 years with a mistaken belief in the existence of the Higgs Boson Particle.

Future generations of scientists and engineers would be kept from learning the truth about the universe and kept from making revolutionary discoveries.

The Vortex Theory reveals that the "mass" of the matter in the universe is not created by particles. Instead, it is being created by distortions in the space surrounding matter.

The mathematics that first revealed the existence of this new vision of the universe was first presented in 2000 in a book entitled: *The End of the Concept of Time*. This same mathematical analysis was then presented as a thesis and was awarded a PhD in 2005 by the Russian Ministry of Education. Eventually it was published in 2013 in two landmark papers by the branch of the Russian Academy of Sciences Journal located in St Petersburg State University, Russia. This unprecedented mathematical analysis reveals a shocking new and revolutionary vision of the five "pieces" of the universe: matter, space, time, energy, and the forces of nature.

This revolutionary vision of the universe has now explained *all* of the great mysteries of the universe, including "Mass"! Now called "The Vortex Theory", this vision of the universe reveals that what have been called particles of matter, such as protons and electrons, are really three dimensional holes in the three dimensional surface of fourth dimensional space. The space surrounding these spherical holes resist deformation creating the characteristics we identify as mass and inertia.

MASS

The above explanation of Gravity reveals that the intervention of an intermediate particle such as the "graviton" is not needed. Because the mysterious force of gravity is now easily explained by the distortions of the holes out of which matter is really constructed; it is also now easily seen that the intervention of an additional "particle" such as the Higgs Boson is not needed to create the characteristic of matter called Mass!

Because science defines mass as *the characteristic of a particle that resists acceleration*, it can now be deduced that the characteristic of a "particle" that resists acceleration is actually being created by the resistance of the internal surface of a 3d hole to oppose the stress created in it by some outside force needed to distort and then accelerate it: and it is this resistance to the distortion of its 3d hole [and the 4d holes within it called quarks] that is responsible for creating the characteristic of matter we call "mass". For example, for 3d holes, observe figure below:

Figure 16.1

Internal 4d volume

3d surface of a hole in space

Surrounding 3d space

Internal pressure of 4d space

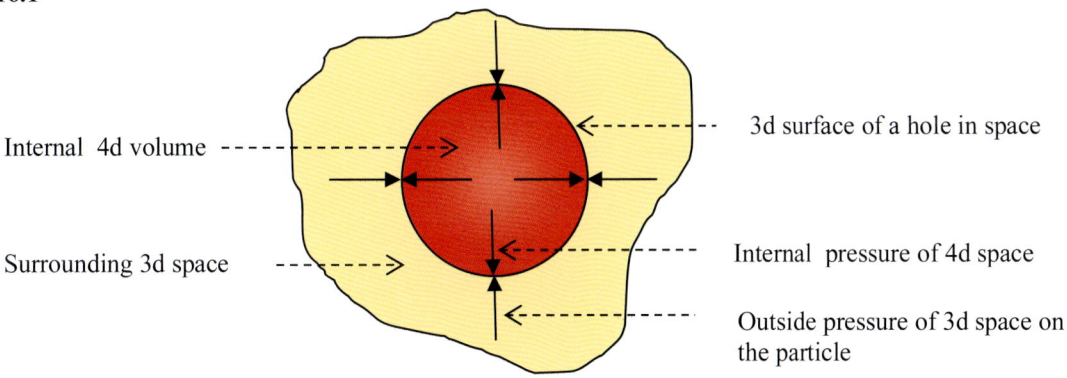

Outside pressure of 3d space on the particle

Figure 16.2

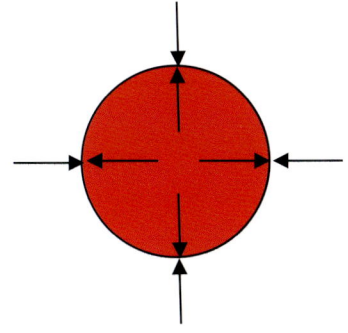

In the above figure, notice how the inward pressure of the external 3d space surrounding the hole, and the internal outward pressure of 4d space within the hole are equal. Consequently, the pressures within and without the hole are in equilibrium. It does not want to move in any direction. It will stay where it is in space unless the outside pressure is lessened on one side.

Also, since the Vortex Theory reveals that quarks [to be discussed shortly] are higher dimensional holes existing within fourth dimensional space, the added difficulty of 3d space distorting this higher dimensional hole is responsible for the quark's seemingly huge mass.

THE GREAT MASS OF QUARKS

The reason why tiny quarks appear so massive comes from the difficulty in 3d space's ability to move them. Because 4d holes cannot be drawn, the analogy of the sphere within the plane analogy must be used:

Note how a 3d sphere has been placed within a 2d hole existing upon a 2d plane. Note too, how each illustration represents a different perspective of the same two objects: the plane and the sphere.

Figure 16.3

Our three dimensional view:

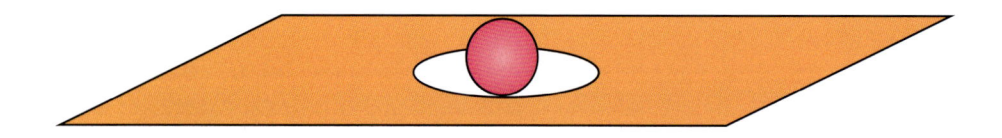

Figure 16.4

Our side view of the 3d hole, the plane, and the 2d hole. Note: from the perspective of the plane, we can see the 2d hole, but not the 3d sphere.

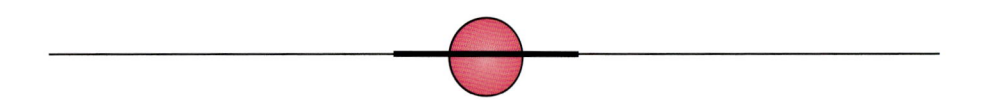

In the second figure, notice how the 2d plane can only intersect and or touch the 3d hole along one single line. Realizing the sphere is actually constructed out of an infinite number of 2d planes allows us to realize that the 2d hole can only interact with the 3d sphere along *one and only one cross-section of the sphere*.

Figure 16.5

The dotted line represents the only place on the sphere that the 2d plane can interact with it.

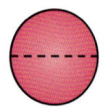

Now, transferring this analogy to a 3d hole and a 4d sphere, we suddenly realize that the 3d hole can only interact along one 3d "plane" of the 4d hole. Consequently, because 3d space can only interact with the much larger 4d hole along only one plane, it becomes much harder to move the 4d hole. Making it [the UP quark or DOWN quark] appear much more massive!

Because the Vortex Theory also reveals that CHARM QUARKS and STRANGE QUARKS are actually 5d holes within UP and DOWN QUARKS, this next layer of quarks are harder still to move, making them even more massive. And finally, since the Vortex Theory reveals that TOP and

BOTTOM QUARKS are 6d holes existing within CHARM and STRANGE QUARKS, these are even harder still to move. Making them seem to be the most "massive" of all.

MOVEMENT OF PHOTONS PASSING GRAVITATIONAL FIELDS

It must also be noted that according to the Vortex Theory, photons are dense regions of space thrown from vortices, expanding and contracting as they travel through 3d space. However, as they pass near "massive objects" such as the sun, they are affected the same way as the holes:

Figure 16.6

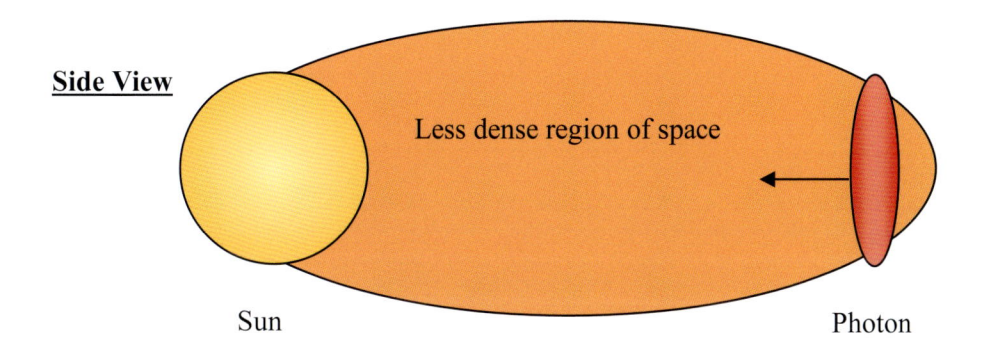

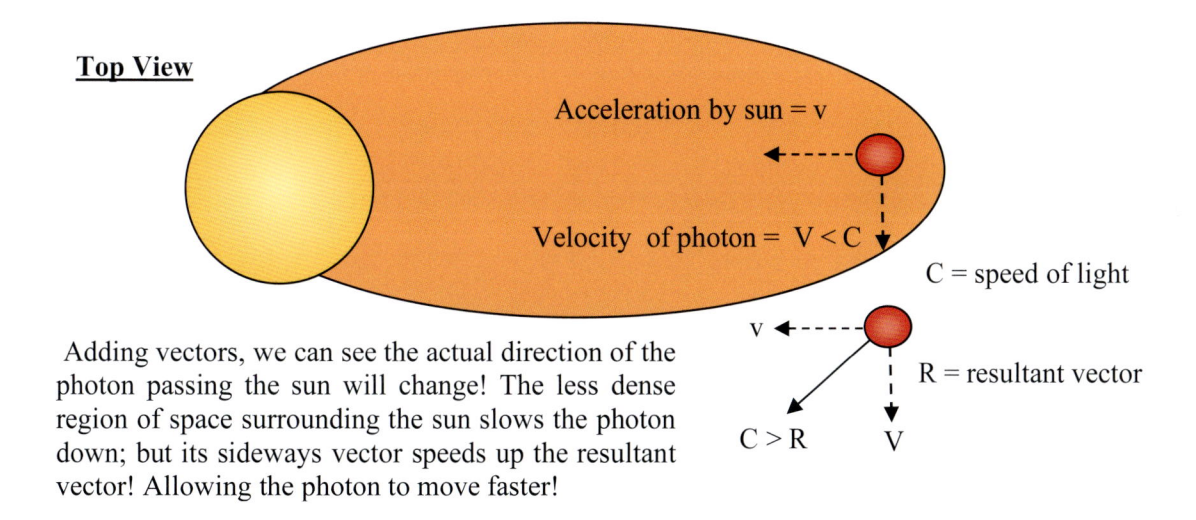

Side View

Less dense region of space

Sun   Photon

Top View

Acceleration by sun = v

Velocity of photon = V < C

C = speed of light

R = resultant vector

C > R

Adding vectors, we can see the actual direction of the photon passing the sun will change! The less dense region of space surrounding the sun slows the photon down; but its sideways vector speeds up the resultant vector! Allowing the photon to move faster!

When a photon travels through a less dense region of space, it slows down because it cannot expand and contract as fast as it can in a denser region of space. However, its acceleration towards the sun by the less dense region of space causes its resultant vector to make it speed up! Seemingly violating the belief that the speed of a photon will always travel at C: speed of light.

As the photon moves through space and encounters a less dense region of space such as that around the sun, again, the side of the photon facing the less dense region of space is distorted slightly in the direction of this less dense region. When this happens, the stress on the photon causes the side facing this less dense region of space to distort towards it. This immediately causes the opposite side to move towards the distorted side, releasing the strain and in effect, accelerating it slightly towards the star. As this continues to happen again and again, the photon changes its direction slightly towards the star. And when a picture is taken of the background of stars during an eclipse of the sun, the light from the stars will appear to move in, towards the sun.

CONCLUSION

The phenomenon of the mass of matter can now be explained as the resistance of the holes matter is made out of to be distorted. The harder it is to distort them, the greater their perceived "mass". Hence, the larger masses of particles containing quarks can be explained by the resistance of the quarks to distortions in their higher dimensional holes sheathed within their three dimensional holes. And finally, the mass of photons can be explained as denser volumes of three dimensional space traveling at the speed of the vortices: C, the speed of light.

Chapter 17
The Terminal Velocity of Atoms Is the Speed of Light

> The maximum speed the atom can move is also a function of the space flowing in the vortex. It is an interesting observation to note that the velocity of the atom itself cannot exceed the speed of the vortex. If it did exceed the speed of the vortex, the vortex flowing in the direction of travel from the proton to the electron could never catch up to the electron; making the speed of light C, the terminal velocity for all the matter of the universe. But there is something even more amazing: The True Speed of Light Cannot be Measured!

Because of a current misconception that space would have to be very dense to allow light to travel at the incredible speed that it does and therefore could not bend or flow - this erroneous deduction must be dispelled.

The way to correct this erroneous way of thinking is to mention an incredible fact that surfaced during the course of these investigations: the speed of light might not be very fast at all! *In fact, how fast it is really moving is unknown and can never be measured.*

This amazing discovery came from the deduction that every instrument (such as clocks, thermometers, etc.) used as a reference to compare other motions is made out of atoms. Because the motions of atoms are functions of the speed of the vortices, and since all of the other motions are created by the speed of the vortices, we are using measuring devices whose own motions are themselves functions of the speed of the vortices. These motions are then being used to measure the speeds of other motions that are also functions of the speed of the vortices. Because there is no other motion independent of the vortices to check the speed of the vortices with, the vortices (and the photons of light emitted from them) might very well be flowing at an actual speed of one meter per second or a billion meters per second and we would never know it.

If the vortices were moving at one meter per second, all other motions would slow down accordingly; since the photons of light emitted from the vortices would also flow at 1m/s, everything would continue to appear as it does now. If the speed of the vortices flowed at a hundred billion meters per second, all other motions would speed up as well; the light emitted from the vortices would flow at one hundred billion meters per second - again making everything continue to appear as it does now.

All we can say is this, "That the vortices are moving at what we call 'the speed of light'." We can compare the speed of one motion to another and we can state that one motion is faster or slower than another - but we cannot tell exactly how fast those motions really are because we cannot tell how fast the speed of light really is!

Because of this amazing fact, the length of a second – which seems very short to us, might really be very long! In fact, every motion in the universe might be so slow that it could have taken a year to write this one sentence! Then again, a second might really be incredibly short. It is possible that it could have only taken a microsecond to write this entire book! Either way, we will never know!

The consequences of this amazing set of circumstances allow us to conclude that no matter how fast or slow the vortices are moving, it will always appear to us and our instruments that they are moving at 186,000 miles per second – the speed of light!

Chapter 18
The Shocking Truth About Time!

> In the first book, The Vortex Theory of Atomic Particles, and the Proof of the Vortex Theory, it was revealed that time, what we call "Time" doesn't exist. Instead, time is a function of motion, a shadow of motion and like a shadow, cannot exist apart from motion.
>
> The phenomenon of time is the result of all the harmonic and sequential atomic, chemical, biological, and astronomical motions creating an orderly sequence of constantly repeating events. These constantly repeating events create a sense of order and harmony in the universe.
>
> However, according to the discoveries of The Vortex Theory, we can now demonstrate that it is the vortices flowing at the speed of light, into and out of the three dimensional holes matter is made out of that are responsible for creating this order and harmony in the universe.

Although it is hard to believe, "time" does not exist and never has. For thousands of years, what man has called "time" is really an illusion being created by motion. Just like the revolving earth was responsible for creating the mistaken belief that the universe was circling this planet, a symphony of harmonic motions is creating the mistaken belief in the illusion called "time"!

The mathematics of the PhD Thesis in the back of Book One reveal that what man calls "time" is actually a function of motion - a phenomenon created by motion, and it only exists within the minds of men. *Time Does Not Exist as a Fundamental Principle of the Universe.* A revelation possessing profound implications!

If the concept of time were merely a philosophical concept and nothing more, its demise would be a significant event - but not necessarily one possessing earth-shaking ramifications. Unfortunately, time is much more than just a philosophical concept. Not only is time an important part of the everyday life of the peoples of the modern era, the belief in time is one of the foundations of 20th Century science.

Destroy the concept of time and everything based upon it is destroyed too. And the extent of the destruction is breathtaking:

Although a fourth dimension of space exists, there is no fourth dimension of "space-time". When Albert Einstein used time to explain the length shrinkage and time dilation effects that take place at near light velocities, he was trying to explain effects that do exist, with a concept that doesn't exist. However, this was not Einstein's fault. He was just another unfortunate victim of the erroneous beliefs of our ancient ancestors.

Like all of the great scientists before him, including Newton, Einstein believed time was a fundamental principle of the universe. He believed this to be true because like everyone else, he was raised from childhood to believe it was true.

Unfortunately, what Albert Einstein and everyone else was taught to believe from childhood was wrong. "Time" is not "real". *Time only exists as a "real phenomenon"*. This fact can be seen in the MATHEMATICAL PROOF at the end of Book 1.

The human race has lived with the concept of time for at least ten thousand years. But until now, nobody has known what it is or how it is created. However, that is all going to change.

Here, at long last, is the explanation of how "time" – or rather the phenomenon of time is created:

Chapter 19
How the Phenomenon of Time Is Created

> Although "time" does not exist as a fundamental principle of the universe, the phenomenon of time does. The phenomenon of time is created by a hierarchy of speeds that begins with the elasticity of space.

The elasticity of space, its ability to bend and flex, is responsible for the fastest speed that space is capable of moving. The elasticity of space is responsible for the velocity of the space flowing into and out of the three dimensional holes of "matter". This speed is not only responsible for the speed of the two vortices of flowing space that join the proton to the electron; it is also responsible for creating the speed of light.

The speed of light appears to be the fastest motion in the universe. This is the key speed, and it sits upon the topmost peak of a pyramid of speeds, however, this speed is only a reference. The speeds of everything in the universe are a function of the speed of the vortices.

The speed of the electron as it moves about the proton is also a function of the speed of the vortices. Since the electron is a hole created within three dimensional space by the flow of one of the vortices, its position on the shell of the atom is determined by the changing position of the vortex. The electron can never move faster than the speed of the vortex that it is attached to. Hence, the speed of the electron as it moves about the proton is a percentage of the speed of the vortices.

Since the proton is also a hole in space which three dimensional space is flowing into, it cannot move faster than the speed of the space flowing into it. Hence its speed is also a function of the speed of the space in the vortices. The same is true for the three dimensional space circulating within the neutron.

Since all atoms are made up of protons, electrons, and neutrons, the speed of the motions of all atoms is again a percentage of the speed of the vortices. The same holds true for molecules.

Since all molecules are made up of atoms, the speeds and the motions of all molecules are percentages of the speeds and motions of the atoms out of which they are made. Hence, a molecule's speed is also a function of the speed of the vortices as well.

And last but not least, at the very end of the hierarchy of motion, is the motion of planets, stars, and galaxies. Because all planets, stars, and galaxies are made out of atoms and molecules, their much slower motions are also functions of the velocity of the space flowing within the vortices. (Note: the above is also true for antimatter atoms and molecules made up of anti-protons, positrons, and anti-neutrons.) This hierarchy of motions now allows us to explain how the phenomenon of time is created.

All harmonic motions associated with the passage of time, including those created within clocks, are functions of, and percentages of, the speed of the three dimensional space flowing within the vortices. Because all the vortices of all the atoms in the universe are flowing at the same rate, the same motions within every atom of every similar element and molecule throughout the universe also take place at the same rate.

It should be noted here that if the vortices were flowing at different speeds at different places in the universe, the speed of the photons of light thrown from these vortices would be variables instead of constants. This would make the light coming from distant suns and galaxies travel at different speeds, creating different spectroscopy frequencies for the same elements. Since this is not seen, the speeds of all the vortices are the same throughout the universe.

These similar motions are responsible for the creation of similar harmonic motions occurring everywhere at the same rate. These similar atomic, chemical, biological, and astronomical motions, no matter where they occur in the universe, create an orderly sequence of constantly repeating events. These constantly repeating events create order and harmony in the universe. This order and harmony is responsible for the creation of the phenomenon of time.

It can now be understood that all similar motions occur everywhere at the same rate not because of some metaphysical quality called "time", but simply because the *dense and flowing space* out of which everything in the universe is made moves at the same rate everywhere!

Because it could also be misconstrued that time does indeed exist and is in fact controlling how fast space can bend and flow, this mistaken idea must be dispelled. Realizing that the "bend-ability" of space is a function of its elastic properties dispels this idea. This "bend-ability" is an important subject because it relates to the "solid-ness" of space.

The topic of the "solid-ness" of space came up when it was postulated that space could not be made of anything because it would have to be extremely dense to allow light to move through it so fast. Consequently, if it was very dense it could not flow – making all of the ideas about *dense and flowing space* false. However, as explained in Chapter 17, this type of reasoning is now revealed to be incorrect.

Chapter 20
Time Dilation Is Finally Explained

> The understanding of how the phenomenon of time is created now allows us to finally understand how the phenomenon of time dilation is created! Including the famous "Twins Paradox".

THE EXPLANATION OF THE FAMOUS TWIN PARADOX

Perhaps the most fascinating phenomenon that exists in the universe today is "Time Dilation".

When an object begins to move faster, from the 20th Century scientific view, its relationship with "time" somehow slows down. This is a real effect possessing astonishing results. The most famous of these effects is the "Twins Paradox":

Say for example when two twin boys were twenty years old, one of them left the Earth for a long voyage on a spaceship to a distant star. This spaceship, traveling at .866C, which is nearly the speed of light (where "C" stands for the speed of light), reached the distant star and immediately turned around and came home.

Because the entire round trip took forty years, when the spaceship returned to Earth, the brother who was now sixty-years old was there to meet him. However, when the spacecraft door opened, instead of him greeting another sixty-year old man, he was amazed to find that his brother had only aged twenty years!

Also, even though the earthly calendar indicated he was gone for forty years, the calendar on the ship indicated he had been away for only twenty years! Amazed, they both realized that somehow, some kind of "time dilation" effect had occurred. Yet neither could explain how it happened. (Neither could anyone else in the world explain how it happened until now!) Here is how it happened.

When the spaceship left the Earth, and began to travel at a near light velocity, the atoms the ship was made of were moving at a much faster velocity than the atoms of which the Earth is made.

This means the electrons, protons, and neutrons of every atom within the ship are moving faster too. And again, just like before, **_even though the vortices connecting every proton and electron continue to flow at precisely the speed of light_**_, from the **point of view within a proton [from PART III], an electron**_, the speeds of the vortices appear to change. From this point of view, one vortex appears to be going much faster than the speed of light while the other appears to be going much slower!

The addition of these two speeds creates a slower round trip time for the space circulating within the vortices. Making the round trip take longer. Making the same round trip take place at what can be called, "THE SLOWER APPARENT VELOCITY." [Note: the same effect occurs for the space circulating within the neutron.]

Also, because the spaceship is moving, space is constantly reconfiguring itself around every tiny three dimensional microscopic hole the ship is made of. As each individual hole moves, the space surrounding it is first bent towards it, and then away from it as it passes by. So even though a proton, electron, or neutron may not be attached to an atom, the space surrounding it ends up moving in two opposite directions, making it reconfigure around the particle at the rate of the slower apparent velocity too. Furthermore, because everything in the universe is constructed out of _dense and flowing space_, including energy, and the "forces of nature", all of their motions, and all of their interrelated motions, will also slow. In fact, the mathematics reveals that all of their motions will slow down exactly, and proportionally to the correspondingly slower apparent velocity of space when they are traveling in the faster moving frame of reference.

For simplicity's sake, the spacecraft was said to be moving at a velocity of .866C. This speed was chosen because at this velocity, the round-trip time of the vortices of every atom within the ship doubles. And when the round trip time doubles, and the reconfiguration of *dense and flowing space* doubles, every motion of every atom within the ship also doubles. Because of the hierarchy of motions described before, when the motions of atoms double, the motions of the objects made out of them double too. Making every motion now take *twice as long* to complete.

Consequently, when the ship was sitting upon the Earth, if it took one second to throw a baseball from one man to another within the ship, when moving at .866C it now takes two seconds: twice as long. Hence, the ball itself now moves at a slower velocity that is directly proportional to the SLOWER APPARENT VELOCITY of the atoms it is constructed out of.

Also, since the atoms in the man's arm are now moving at the slower apparent velocity of space, it cannot move as fast as it could when the ship was motionless. Therefore, the man cannot throw the ball as fast as he could when the spaceship was not moving. However, since the motions of every atom in the ship have slowed down proportionally to the slower apparent velocity of space, NOTHING appears to have changed! If the man was capable of throwing the ball at a speed of one hundred miles an hour when the ship was motionless, to all of those within the ship, the ball still appears to be moving at a speed of one hundred miles an hour! From the perspective of all those within the ship, everything "is as it always was". But this is just an illusion.

In this faster moving frame of reference, the motions of everything have now slowed down proportionally to the motions of everything else. This means that EVERY motion, including all motions within all assorted types and kinds of clocks, now takes place at a slower rate than it did when it was at rest in reference to the Earth.

Hence, a clock attached to a wall in the interior of the spaceship now runs slower in reference to a clock on the Earth. It runs slower because every motion of every atom within the clock now takes twice as long as it did before. When it used to tick off two seconds, it now ticks off only one second. One hour becomes a half-hour, and 24 hours turns into 12!

[If an atomic clock is launched into space aboard this spaceship, orbits the Earth at a high rate of speed, and is then brought back down to the surface; comparing its time to the time on another atomic clock will make it seem as if this clock has run slower (which it has). However, it was not "time" that was the cause of this "Time Dilation" effect. Instead, it was the slower apparent velocity of *dense and flowing space* that was the culprit. Allowing us to see the real cause of this time dilation experiment that is supposedly used as a proof of Einstein's Theory of Relativity!]

Returning our attention to the spacecraft, because all motions have slowed down proportionally to all other motions, the biological processes within the traveler's body have slowed down too! This means that in addition to the motions of the atoms and molecules within the traveler's body, his very thought processes will also have slowed down proportionally to the slower motions of the atoms within his body. Since the electric current and chemical reactions within his body have slowed down, his thoughts and perceptions have slowed down, making it seem to him as if nothing within the ship is any different from what it was before!

Chapter 21
The True Vision of the Universe

The true vision of the universe is a shocking vision unlike anything anybody has ever seen before!

The truth about time is the final piece of the puzzle. It is the union of these "five pieces" of the universe that completes the picture and finally allows us to see the ultimate vision of the universe:

[When I first saw the true vision of the universe – the vision that no man has ever seen before – I was awestruck! Thunderstruck!]

This true vision of the universe is the greatest scientific discovery ever made! It is a shocker! A blockbuster!

Stated simply, there are no separate "parts" to the universe! There is only space and motion! Everything that exists is part of the same thing! EVERYTHING IS "ONE"!

The only thing that exists in the physical universe is the substance of which space is made. Space is not "space" at all. Space is made of something that is totally unique from our point of view. It is constructed out of at least seven dimensions [as will be revealed in Book 3, The Quark Theory], and it can both bend and flow.

This substance is in motion. It is moving in two opposite directions at once: on a massive scale, it is expanding outward at an incredible speed, carrying the Galaxies with it. But on the microscopic scale, like the eddy currents in a river, it is flowing backwards at an equally incredible speed (from our point of view)!

Both matter and energy are created out of space. The "particles" of matter - protons, electrons, and neutrons - [and a host of other "particles"] are not particles at all. They are three dimensional holes existing upon the surface of fourth dimensional space. The substance of which space is made flows into and out of these holes.

The hole creates a particle effect, while the denser or less dense space surrounding the hole creates a wave effect.

Energy is also created out of dense and flowing space. What we call energy is merely denser regions of space called photons that expand and contract as they move. This dense region - or photon - creates a particle effect, while the expansion and contraction of the space it passes through creates a wave effect.

Time does not exist; the phenomenon of time is created throughout the universe by the uniform flow of microscopic space into and out of the three dimensional holes of matter.

There is NO universal "force" of nature! All forces are created out of unique configurations of space.

Chapter 22
At Last – The Magic Cipher!

> A cipher is a key that allows a secret agent to unlock secret communications. The same is true for the discoveries in the first half of this book. The explanation of the five pieces of the universe: matter, space, time, energy, and the forces of nature have all been explained using the principles of bent and flowing space. But this is not the end, just the beginning.
>
> The relationship between these five pieces of the universe now allows us to explain a whole host of previously unexplained mysteries, some of these now will be presented in PART II of this book.
>
> It should also be stated that with the explanation of seven dimensions of space, the many and great mysteries of the quark theory will also be explained in Book 3.

At last, indeed! What we have waited for, for hundreds of years is here at last! The magic cipher that allows us to explain everything in the universe! What can only be called ***THE ULTIMATE VISION OF THE UNIVERSE!***

The knowledge sought from the beginning of civilization by the greatest minds of the greatest philosophers and scientists who have ever lived. Men and women who spent their lives in search of the truth: the ultimate vision of the universe, the great ones such as Socrates, Aristotle, Plato, Hermes of Thebes, Copernicus, Galileo, Maxwell, Newton, Madam Curie, Einstein…and the list goes on and on.

But as said before, this is just the beginning. This "Magic Cipher" allows us to explain mysteries of the universe that nobody could explain before. It also allows us to correct mistakes that were made by past scientists, making incorrect explanations that we can now identify as false. A great and wonderful window has now opened up upon the universe: nothing will ever be the same again!

PART II
THE GREAT MYSTERIES OF THE UNIVERSE FINALLY EXPLAINED

The following explanations vary in length; some only take a few lines, while others take a few pages. I have spent my life in pursuit of the answers you are about to discover. I hope you enjoy them. Some create visions of enlightenment, some invoke fear. If so, remember, that although it is hard to believe, and what most of us fail to recall, is that we exist almost like microbes upon the surface of a vast spinning sphere of matter, hurling over 65 thousand miles per hour through a vast dark void of empty space, occasionally lit up by burning spheres of gas many hundreds of thousands of miles in diameter. A shocking vision that invokes both fear and awe!

#1 The Explanation of the Particle and Wave Theory of Light...

Although quantum mechanics presently believes that the particle and wave theory of light has already been explained, it is greatly mistaken!

Although the photon was already explained, because it is one of the great discoveries of the Vortex Theory, it is listed in this section with the great discoveries of science. Also, a brief synopsis is given here to prepare the reader for the explanation of the double slit interference patterns.

So, to reiterate, photons of energy are condensed packets of three dimensional space that displace the surrounding three dimensional space outward, creating a spherical region of dense space that surrounds the photon. The condensed region of space within that we call the photon, creates the *particle effect*, while the region of denser space surrounding it combined with the photon's expansions and contractions [as explained further along] creates its *wave effects*.

The photon can now be defined as a packet of condensed three dimensional space, expanding and contracting in a long tubular shape perpendicular to its direction of travel. The rate of its expansion and contraction - or frequency - is a direct function of the volume of condensed space within the photon. And just as before, using the principles of "less dense and flowing space", another one of science's greatest mysteries - the particle and wave effect of light - is easily explained!

#2 The Explanation of The Particle and Wave Theory of Matter...

Although according to the generally accepted quantum mechanics theory, a particle and wave theory of matter has already been explained: again, this is not true!

THE PHOTON

As with the photon, although the particle and wave theory of matter was already explained earlier in this book, because it is one of the great discoveries of the Vortex Theory, it is listed in this section with other great discoveries of science. A brief synopsis is given here to prepare the reader for the explanation of the double slit interference patterns coming next.

THE ELECTRON

Just as the proton is a three dimensional hole bent into fourth dimensional space, the electron is a three dimensional hole bent outward, out of fourth dimensional space. A whirling vortex of three dimensional space passes through the electron from fourth dimensional space and back into three dimensional space. This outward flow of three dimensional space creates the electron's electrostatic charge.

As this charge of three dimensional space pushes outward, out of the electron, it pushes the surrounding three dimensional space outward, surrounding the electron with a massive region of *dense* space. Although this region is massive in relation to the size of the electron, it is much smaller than the size of the less dense region of space surrounding the proton.

And again, just like the region of less dense space that surrounds the proton, the region of dense space that surrounds the electron is a function of the three dimensional space the hole is traveling through. This means that as the hole moves through the three dimensional space of our universe, the space the hole is passing through bends outwards away from the hole as it approaches, and then back inward as it passes by. This dense region can be analogized as a "hill" or a "bump" in space.

Consequently, even though this massive sphere of dense space constantly surrounds the electron and seems to travel with it, *it is the region of space the hole is passing through that is creating this sphere.*

PROTONS AND NEUTRONS

The "particle" effect of matter is created by the three dimensional hole in space, while the density of the space surrounding the hole creates the "wave" effect of matter. For a proton, this region is created when the surrounding space is pulled into the hole, creating in effect a "valley" or a depression in space.

This region of less dense space surrounds both the proton and the neutron, just like the denser region of space surrounds the electron. When considered which bend would be the best candidate for creating the source of gravity, it was realized that it had to be the bend surrounded by the less dense region of space. Within this less dense region, space would be "stretched inward". This less dense region would make the space surrounding other holes stretch towards it, making the holes move in that direction too.

Because it was previously concluded that protons and neutrons were responsible for creating the less dense space surrounding stars, to create this less dense region, three dimensional space is flowing into the proton, and out of the electron.

Even though this idea is contrary to the present scientific beliefs of how the lines of electrostatic forces are pointing, it must be remembered that the direction of the positive and negative electrostatic forces were originally and arbitrarily assigned by Benjamin Franklin. Nobody knows for sure just which direction is which. [It should also be noted, that the mathematical proof of this theory will work no matter what direction space flows.]

The particle and wave characteristics of matter are now easily explained. What was once perceived to be a very complicated idea is now rendered amazingly simple.

Although many people nowadays understand that photons of energy possess both particle and wave characteristics, it seems to defy logic that matter should behave as both a particle and a wave too - and yet it does. As strange as it seems, matter possesses characteristics that make it behave both like a rock held in the hand, and sometimes like the waves it creates upon the surface of a pond after it is thrown in.

In the past there was no explanation for this dual phenomenon, but now there is: while the "holes" in three dimensional space are responsible for the particle effects of matter, the less dense (or denser) regions of space surrounding them are responsible for the "wave" effects of matter. Clearing up one of the great mysteries of science!

#3 The Explanation of The Double Slit Interference Patterns Created by Light and Matter

> In 1801, an English physicist named Thomas Young performed an experiment where light went through a double silt creating an interference pattern of light and dark lines on the wall behind the slits. Young believed this was proof of the wave theory of light. However, and shockingly so, it was later discovered that single photons going through the slits and hitting photographic paper behind, over time and as the more and more photons hit the photographic paper, they created the exact same series of light and dark lines!
>
> This seemed contrary to logic. Logic suggests that only two lines would be formed [one behind each slit]. It was and is [until now] a great mystery. Even the great Richard P. Feynman, winner of the Nobel Prize in 1965, was greatly disturbed by these results. And when he received his PhD in physics and called his father to give him the good news, his father asked him if he could answer one question he had about physics. When a confident and proud Richard said, "Sure ask away", his father said, "Can you explain how single photons going through Young's two slit apparatus create the same number of lines as when they travel through en mass?" Unfortunately, and after an embarrassing pause, Richard later said he could only answer, "No!" And when he did, his father hung up the phone. To which Feynman later admitted to having sat down and cried; because after 8 years of education, he could not please his father, as he so desperately wanted to.
>
> And this problem has continued to this day. Because until the discovery of the Vortex Theory, nobody could answer this one question, now it is easily explained!

The great mystery of how a single photon or electron can create the wave like interference pattern in the famous twin slit experiment is easily explained by the Vortex Theory. In fact, this mystery is so easily explained, that it now becomes almost a trivial subject!

The wave-like interference pattern made by projecting either light or matter through the twin slits *is created by the dense or less dense regions of three dimensional space surrounding photons, electrons, and atoms.* As far as dense space goes, in comparison to the size of the photon or electron, the dense region of space that surrounds each of them is massive.

Figure 3.1

Note: in relation to the size of the photon or the electron, the region of dense space that surrounds the photon or electron is so massive it is impossible to draw the proportionate sizes.

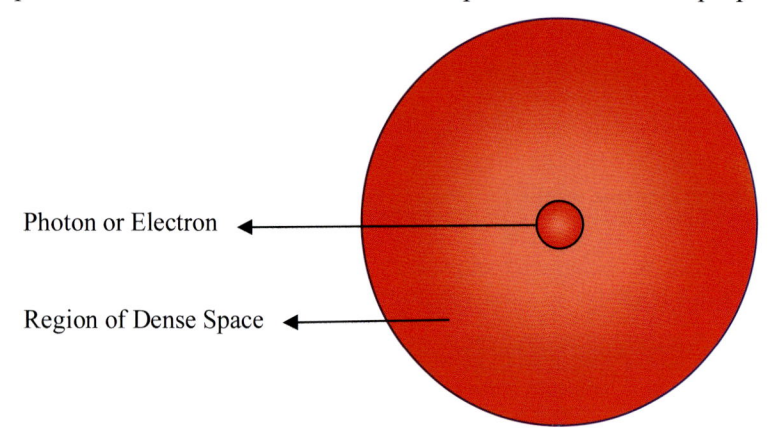

Photon or Electron

Region of Dense Space

This region is so massive, that when a photon or electron approaches the twin slits, its denser region arrives first.

Figure 3.2

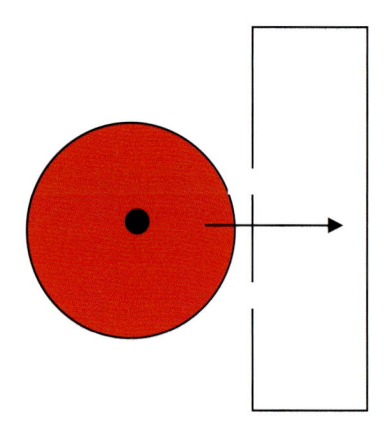

Because the dense region arrives first, it begins to go through the slits before the photon or electron [black dot]

Figure 3.3

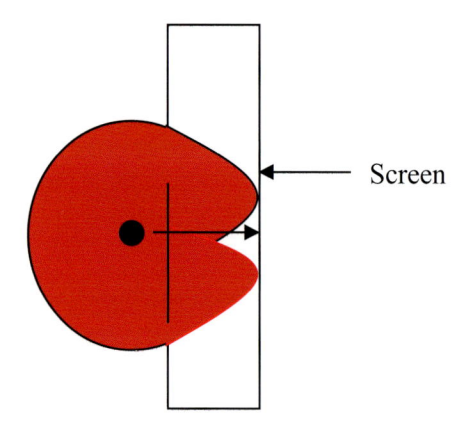

Screen

As this denser region moves through both slits simultaneously, the two waves interfere with each other and begin to create the invisible wave patterns of dense and less dense space on the screen.

Figure 3.4

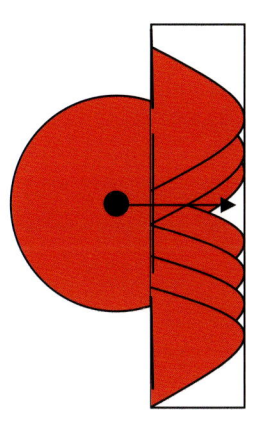

When the photon or electron finally moves through one of the slits, it moves into a region of space that is alternately dense, less dense, dense, less dense, etc. Consequently, the surface of the photon or electron is distorted towards *one of the less dense regions*, avoids the denser regions, and travels in the direction of this less dense region, striking the screen.

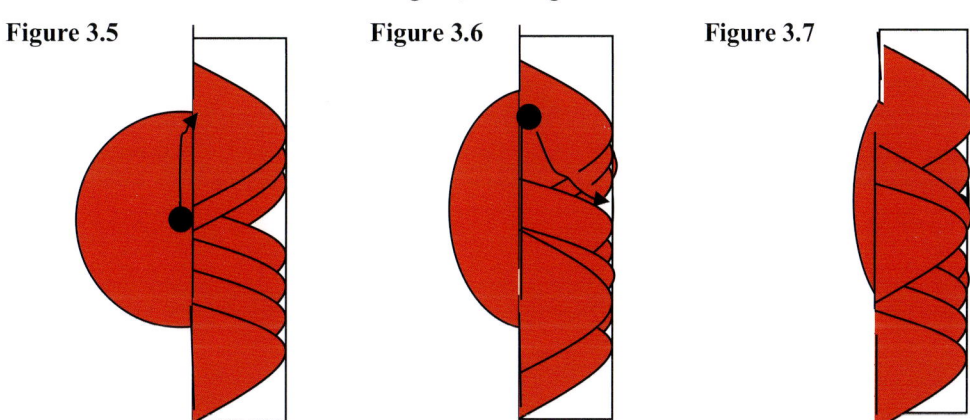

Figure 3.5   **Figure 3.6**   **Figure 3.7**

Note, in Figure 3.6, as the photon or electron passes though one of the twin slits, note how the dense region behind the slit has not yet passed all the way through. This follows through of the sphere of dense space behind the photon or electron continues the interference pattern until the photon or electron strikes the screen. Note too, this picture shows the photon or electron going through the top slit. If the region of dense space were oriented more towards the bottom slit, it would have gone through that one.

When this same situation happens over and over again, it is their collective statistical behavior that sends them to the less dense regions, avoiding the denser regions, and creates the light and dark strips seen on the screen.

Figure 3.8

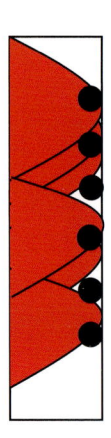

Because it can now be seen how a single electron or photon can create the interference pattern, some important misunderstandings can now be laid to rest.

#1. Perhaps the most important misunderstanding is the mistaken idea that somehow the electron or photon was thought to go through both slits at the same time. As can now be seen, this is a mistake.

#2. Equally important is the idea that photons of light somehow interfere with each other, "cancel each other out" creating the dark spots on the screen. This is also a mistaken idea. Light waves do not "cancel each other out". Instead, it can now be seen how electrons and photons avoid the denser regions and are attracted to the less dense regions. [Note, both the shape of the photon and the three dimensional hole of the electron are distorted in the direction of the less dense regions of space causing them to move in those directions.]

#4 The Explanation Of ½ Spin of Particles

One of the most fascinating characteristics of sub-atomic particles is what is called ½ spin. They all seem to be spinning in a way that is called UP or DOWN by physicists. Until now, this strange characteristic could be observed but not explained. But all that is ended now…

In 1921, two scientists, Otto Stern and Walter Gerlach performed an experiment that showed the "quantized" electron spin occurred in only two orientations. (Called spin "up" and spin "down".)

The experiment was made using a beam of silver atoms that were heated in an oven, then shot through a non-uniform magnetic field, and impacted upon a photographic plate. Because the outer electron of the silver atom is in effect shielded by the other 46 electrons, the silver atom allowed them to study the magnetic properties of a single electron. However, because the outer electron has a "zero orbital angular momentum", no such interactions between the silver atoms and the magnetic field were expected. What Stern and Gerlach expected to see was a "continuous smear" on the photographic plate. But this was not to be.

When the silver atoms were directed through the non-uniform magnetic field, instead of seeing a "continuous smear" upon the plate, Stern and Gerlach saw that the beam was split into two separate parts. This split indicated that there were just two possible orientations for the magnetic moment of the electron.

This result seemed impossible, because how can the outer electron of the silver atom obtain a magnetic moment if it has no angular momentum and hence cannot form a current loop that creates a magnetic moment?

The problem was (supposedly) solved four years later when Samuel Goudsmit and George Uhlenbeck proposed that the electron must possess an intrinsic angular momentum. Today this intrinsic angular momentum is called electron spin: (or ½ spin).

But instead of solving the problem, ½ spin only deepens the problem. Because how can the electron only spin in one of two possible orientations? (Again, called spin "up" and spin "down".) Furthermore, how can an electron, or for that matter a proton, neutron, neutrino or any other 'particle" that also possesses ½, spin, spin in only one of two possible orientations? It doesn't seem to make any sense! It is still one of the greatest and most perplexing mysteries in all of today's science!

But no longer! Using the Vortex Theory, ½ spin is easily explained. In fact, it is so easily explained that the children of future generations will be able to explain it to their grandparents!

According to the Vortex Theory, protons, electrons, and neutrons (along with a great number of other "particles") are three dimensional holes existing upon the surface of fourth dimensional space. Because each higher dimension is at right angles to all lower dimensions, fourth dimensional space is not only at right angles to three dimensional space, it is at right angles to all three lower dimensions simultaneously.

Because fourth dimensional space is at right angles to three dimensional space, a three dimensional hole that allows entrance into higher dimensional space is at right angles to fourth-dimensional space. Therefore, its three dimensional surface forms a ninety-degree angle to a fourth dimensional axis of rotation that extends into fourth dimensional space.

Because a fourth dimensional axis of rotation is impossible to draw, a two dimensional to three dimensional representation is shown below:

Figure 4.1 Three dimensional view:

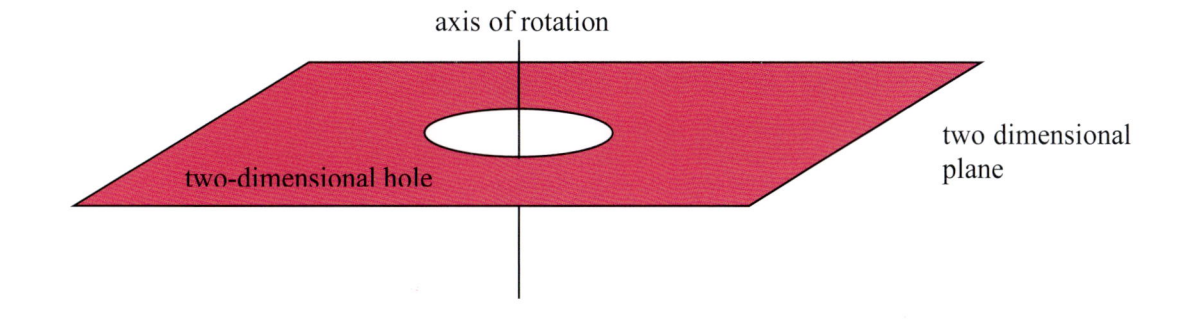

Figure 4.2

Top view:

Note how the two dimensional hole can only be rotated clockwise or counterclockwise!

Note in Figure 4.1 how the two dimensional hole's three dimensional axis of rotation, that extends through its center and into three dimensional space, is at right angles to the surface of the two dimensional hole. Note too, how the hole can only rotate – or spin – in only one of two possible ways, clockwise or counter-clockwise! Even though the above illustration is a two dimensional to three dimensional illustration, the exact same situation – though impossible to draw – is happening between three dimensional and fourth dimensional space.

Because a three dimensional hole possesses a fourth dimensional axis of rotation, it can only rotate clockwise or counterclockwise around its fourth dimensional axis. Hence, it can only rotate in one of two possible spin orientations that have been designated "UP" or "DOWN"!

To better illustrate how only these two rotations can occur, take a piece of paper and draw two lines through the center of the paper at ninety-degree angles to each other forming a large cross. Now label one line "X" and the other line "Y" (for the X, and Y, axis on the Cartesian co-ordinate system). Now take a pin and stick it through the piece of paper at the point where the lines intersect. The pin represents the "Z" axis of the Cartesian co-ordinate system. Because the piece of paper can be thought of as a plane consisting of two dimensions, notice how the pin (that represents the third-dimension) forms a ninety-degree angle to the other two dimensions simultaneously. Now rotate the pin back and forth between your thumb and forefinger.

As you rotate the pin, notice how the pin, that is at a ninety-degree angle to each of the two lower dimensions simultaneously, can only be rotated in one of two possible spin orientations, clockwise or counterclockwise!

Again, although this is a two dimensional to three dimensional analogy, the same exact situation is happening between three dimensional space and fourth dimensional space. The fourth dimensional axis of rotation that is at right angles to all three lower dimensions simultaneously can only rotate in one of two possible spin orientations, clockwise or counterclockwise: "up" and "down".

Consequently, it can now be stated that 1/2 spin is created when a three- dimensional hole or a "particle" such as the electron that is bent into fourth dimensional space, can only rotate clockwise or counterclockwise around its fourth dimensional axis of rotation. Because of this axis of rotation, the hole can only spin in a clockwise or counterclockwise direction about the axis, creating only two possible spin states, clearing up another one of the great mysteries of science.

#5 Explanation of Inertia

> Inertia is defined as the property of matter that causes it to resist any change in its motion. Thus, a body at rest remains at rest unless it is acted upon by an external force and a body in motion continues to move at a constant speed in a straight line unless acted upon by an external force.

Most people could care less about inertia unless one is a baseball player trying to hit a ball into the outfield, or the kicker on a football team trying to put the ball into the end-zone! What it takes to make something move fast and far is not a necessity for most of us in our everyday life. But it is important in the world of science. Especially as we will soon see in the explanation of Sir Isaac Newton's famous three laws of motion.

So, what creates inertia?

Inertia is created by stresses upon the inward surface of the holes matter is made out of. For example, in Figure 5.1 below, the pressures on the internal surface of the proton and the external pressure of space are equal.

Figure 5.1

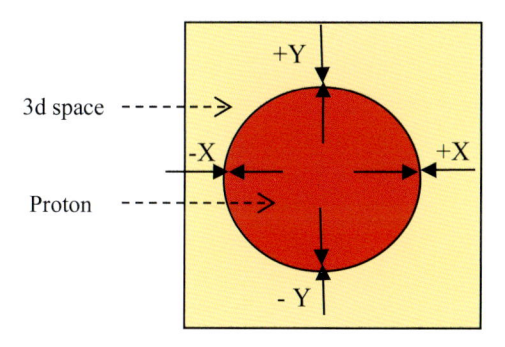

Because the inner and outward pressures are all the same, the proton has no inclination to move in any direction. However, if an external force strikes the proton, the forces are no longer in equilibrium.

Figure 5.2

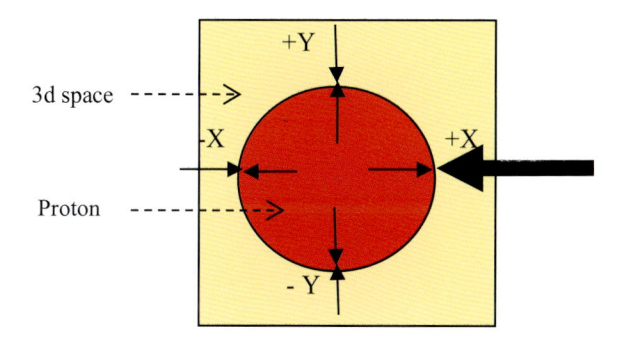

The instant the force strikes the proton, the +X side of the proton is bent in towards the –X side: Figure 5.3. At the same instant, a THREE DIMENSIONAL wave starting at the +X side travels around the inside of the hole along *both* the X axis and the Z axis [not drawn] of the now distorted sphere towards the –X side pushing outward the sphere in the –X direction Figure 5.4:

Figure 5.3   **Figure 5.4**

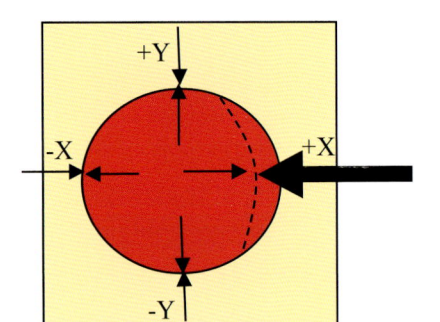

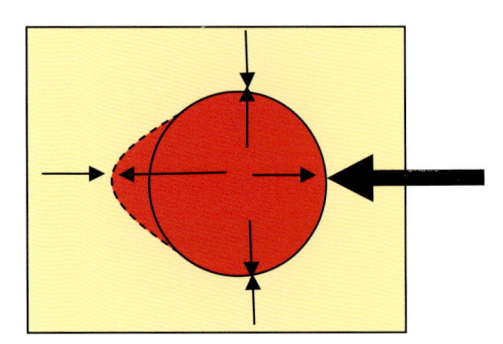

This wave would seem to move the hole as seen in Figure 5.5 and then stop; however, although the hole would like to return to its spherical shape as seen in Figure 5.5, it cannot. Instead, the stress upon the inside surface of the hole causes it to want to straighten out: see Figure 5.6; forcing the back side of the hole to move forwards towards the –X side. However, the outside three dimensional space surrounding the hole is now pushed outward in the –X direction. This relieves the outside pressure upon the hole causing the inside to keep pushing outward into it: Figure 5.6. This makes the back side of the hole [the +X side], continue to try to move forward towards the –X side; and as the outward distortion of the hole towards –X side continues to push outwards into the space in front of it, and the back side of the hole continues to move inward; the proton continues to move onward in the –X direction of the of the Cartesian Coordinate System unless it encounters some other force. Validating Newton's first law of motion [see next section].

Figure 5.5   **Figure 5.6** ------> Internal stress
                                              ------> Internal stress

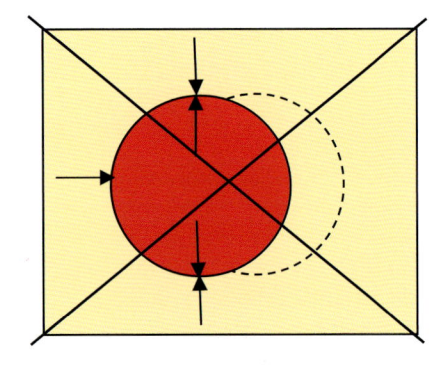

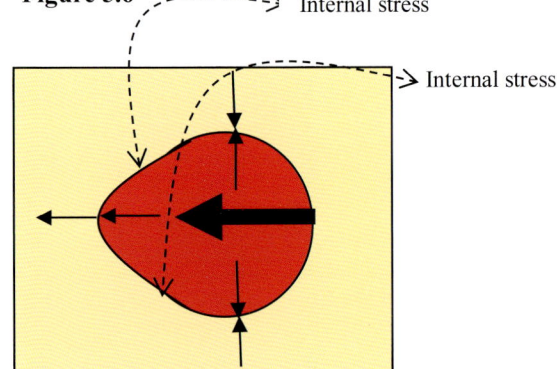

Resolving the conflict between inertial mass and gravitational mass…

One of the famous curiosities of physics was [and to some still is] the great conflict that occurred a century ago between the idea of inertial mass, usually designated as m_i and that of gravitational mass m_g.

For those who don't know about the problem, gravitational mass is defined as that which interacts with a "gravitational field" to give a body its apparent weight; while inertial mass is defined as the product of a body's volume/density which will give the resulting momentum and kinetic energy of the body. Or, in effect, two different definitions of mass!

In an effort to measure the two masses and see if there were any differences, around the year 1900, the distinguished Hungarian physicist Roland Eötvös measured the two using extremely sensitive instruments. He concluded they were the same value to an accuracy of many decimals. However, the conflict still remains due to *Mach's Principle*.

Mach's Principle asserts that the inertial effects of mass are not innate in a body, but instead, arise from its relation to the totality of all the other masses in the universe – (the center of mass of the universe). Because the center of mass of the universe would be so far away from the Earth, and its gravitational effects would be minimal at best, today, this principle does not seem logical. However, because nobody knew how the inertial mass of an object was created, or how big the universe actually was, or where the center of mass was located; both sides of this argument remained valid possibilities that could not be dismissed. But now, it can finally be cleared up.

As we have seen in this study, the mass of an object [a hole such as a proton] is a function of the inward and outward pressures upon a hole that resist distortion.

Also, because this study reveals that the mass of the universe exists within the higher dimensional space of the interior and not upon the three dimensional space of the surface, this entire argument is now ridiculous.

Hence, even though the mass of a hole is a function of the surrounding three dimensional space's ability to distort its surface, its mass is a function of its resistance to distortion.

Because inertia is a function of mass, the definition of mass also applies to inertia: the inertia of a mass can be defined as the property of the space the three dimensional or fourth dimensional hole that resists the initial attempt to be bent and create the internal wave within it. Since the density of space defines how easy or resilient it is to being bent, inertia is ultimately a function of the "bendability" of the space creating the three dimensional or fourth dimensional surface of the hole.

Using the above definition, *inertia can be defined as an effect created by the surface of a three, or fourth dimensional hole's initial resistance to being distorted.*

#6 The Explanation of Newton's Three Laws of Motion

> After 300 years of triumphant use, these three great laws of engineering and physics are no longer observations without explanations. Before they were explained as axioms: truisms that have no proof yet must be accepted as truths. Today that is all ended!

Historical Context

In 1687 Isaac Newton published his immortal *Principia Mathematica* and the world has never been the same again. This great work, perhaps the greatest scientific work ever published, contains amongst other discoveries his three laws of motion.

Although these laws are called axioms because they cannot be proven, but can only be postulated, and then tested by experiment, scientists and engineers have nevertheless used these three laws of motion successfully for over three hundred years. In fact, *all* of the world's great civil and mechanical engineering feats have been accomplished using these laws; and many of the great scientific discoveries in astronomy and physics could not have been made without the use of these laws, especially the second law, F = ma. And yet, until now, these three most important and vital laws of the universe have been nothing more than observations without explanations. But this 300-year failure has just been reversed.

Using the principles of the Vortex Theory, these laws are so easily explained, that in the words of the Immortal Detective, they would be "…Elementary!"

THE THREE LAWS…

First law - a body at rest remains at rest, and a body in motion continues to move at a constant speed along a straight line unless the body is acted upon, in either case, by an unbalanced force.

Second law - an unbalanced force acting on a body causes the body to accelerate in the direction of the force, and the acceleration is directly proportional to the unbalanced force and inversely proportional to the mass of the body (F = ma) or force equals mass times acceleration.

Third law - for every action or force there is an equal and opposite reaction or force.

THE EXPLANATION OF THE FIRST LAW…

First law: a body at rest remains at rest, and a body in motion continues to move at a constant speed along a straight line unless the body is acted upon, in either case, by an unbalanced force.

Luckily, in the previous explanation of inertia, this first law of Newton was explained. The key to understanding this first law was found in the "continuous wave" created within the surface of a three dimensional or fourth dimensional hole as it moves through space.

Once this wave is set into motion there is nothing to stop it unless it encounters another region of bent space that bends it in another direction, or unless it suffers a collision with other matter or energy.

Although the distortion in the space surrounding the hole can be called a "wave", it does not create an Aether wind. An Aether wind would only be created if matter were made of solid particles. Such particles would "blow" space to either side of it as it moves, creating a wind in space that the Michelson Morley Experiment would have detected.

The Second Law…

Newton's second law states that an unbalanced force acting on a body causes the body to accelerate in the direction of the force, and the acceleration is directly proportional to the unbalanced force and inversely proportional to the mass of the body, or (F = ma).

Wow! What a big usage of words for something we can now describe so simply! It goes like this because the formula F = ma, (force equals mass multiplied by acceleration), it simplifies the explanation of Newton's second law by stating that force is directly proportional to both mass and acceleration. Or as we have just seen, mass equals the number of holes present; and acceleration equals the amount each is distorted!

Figure 6.1

In the figure below it can be seen that there are 10 protons sitting on a plate. However, note how none of them are distorted. Hence, we have no force on the gold bar. But this amount of "particles" = "M", Mass, in the famous formula.

Figure 6.2

In the figure below it can be seen that we now have 18 protons: hence we have more MASS. But still no force because none of them are distorted downward towards the gold bar.

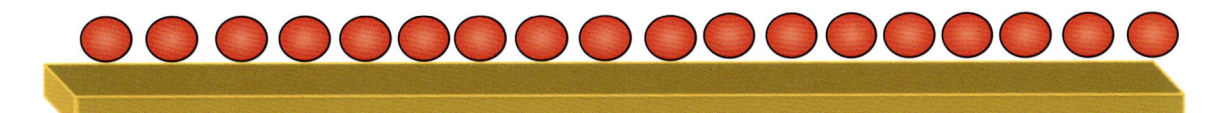

Figure 6.3

Now, in this figure, it can be seen that the protons are all bent downward: accelerated in the direction of the gold bar. This places a pressure, or force, upon the bar as can be seen by the black arrow. These distortions are the acceleration "A" in Newton's formula.

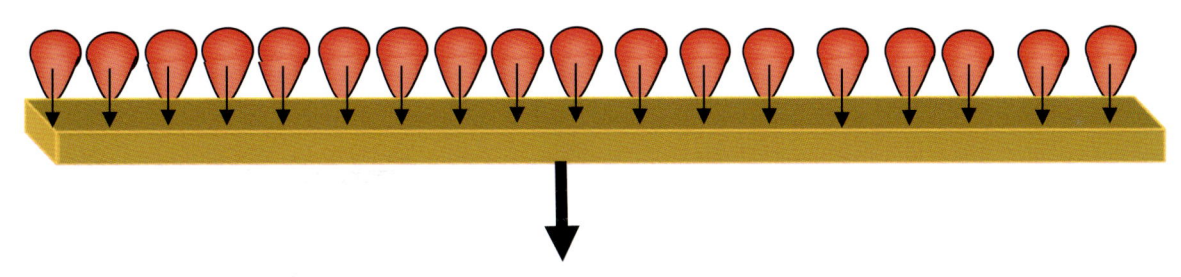

Hence, we can now use pictures to explain this famous formula. Notice too, how all of the small arrows add up to being a force of one big arrow.

THE THIRD LAW…

Third law - for every action or force there is an equal and opposite reaction or force.

Every time I see this law, I wonder if Newton played billiards. Because if he did, he would have seen the actions and reactions of the balls on the table and realized that something wonderful was happening here.

It is too bad he did not know that matter was made of holes in space, and that acceleration was caused by distortions in their surfaces, because if he did I believe he would have easily guessed the truth: that it is the transfer of these distortions from one object to another that is responsible for his famous third law.

This law is the result of the waves within the holes in one object being transferred to the holes in another object – and then being transferred back.

For example, when a man throws a baseball, the acceleration of his arm creates waves within the holes his arm is made of and waves within the holes making up the ball he is holding onto. When he lets the ball go, the waves are transferred throughout the holes within his body, through his feet, and down into the ground he is standing upon. Because the earth contains an enormous amount of holes in comparison to those within the man's body, he cannot move the earth away from his feet, so they are reflected back into his body allowing him to remain standing.

However, if the man is in space when he throws the ball, when the waves are transferred throughout his body there is no longer a solid earth to reflect them back. Hence, as the waves transfer throughout the holes within his body into the holes, he begins to rotate.

Another example of this reverse wave transference around the holes that matter is made of, can be seen in a billiard ball hitting a cushion on one of the rails of a billiard table. When the billiard ball hits the cushion, the combined amplitude of all the waves surrounding the holes within the ball are transferred to the holes the cushion is made of, and then from the cushion, back into the holes the ball is made of. If the ball hits head on, its direction is reversed because almost all of the amplitude of the wave is transferred directly back into the ball (assuming losses due to friction and the slight distortion of the table are negligible).

But if the ball hits at an angle, it reflects at the same angle because only a percentage of the amplitude of the wave was transferred. (This percentage is a value equal to the amplitude multiplied by the sine of the angle of incidence while allowing for losses of friction and the distortion of the table.)

#7 Momentum

> The subjects of *Momentum and the Conservation of Momentum* flow directly out of Newton's laws. Momentum, $P = MV$ [momentum equals Mass (amount of matter) multiplied by Velocity] is created by the amount of matter, multiplied by the amount of the distortion of the holes matter is made of.

When an object is in motion, the sum of the collection of all of the waves within all of the individual holes that make up the matter an object is made out of is responsible for creating its *Momentum*.

When a moving collection of holes collides with another collection of holes [like a pool ball hitting another pool ball], some or all of the waves surrounding the moving set of holes is transferred to the other set of holes. This transference is responsible for the law called the *Conservation of Momentum*.

#8 The Explanation of The Conservation of Momentum

> Unbeknownst to most viewers, the conservation of momentum is seen every day in every sporting event on television. When the man kicks a football or hits a baseball with a bat, it is not the man, but the law of the conservation of momentum that decides the outcome of the play! However, none of the athletes in the game, or the professors in the universities where they play, knew how angular momentum was created until now!

It was decided to explain the conservation of momentum after the explanation of Newton's three laws and the Conservation of Angular Momentum to simplify its explanation. However, because hindsight is always 20/20, this order of explanation now seems unimportant. Because using the principles of the Vortex Theory, the explanation of the conservation of momentum is now so simple to explain, it almost seems trivial!

The transfer of part or all of the internal waves within the surfaces of the three and fourth dimensional holes of one object to another object, is the explanation of the Conservation of Momentum.

The reason why the momentum formula ($P = MV$) works is a result of the "number of holes" present and their velocity [note, mass is only proportional to the number of holes].

According to conventional wisdom, when a large mass (m) is moving at a certain velocity (v), it possesses a certain amount of momentum (P). If the mass increases the momentum increases, and if the velocity increases the momentum also increases.

But when seen by the Vortex Theory, there is another explanation for this equation.

Because a mass is roughly proportional to the number of holes an object possesses, the larger the number of holes, the more waves there are, and the larger the amplitude of the collective wave: increases (P).

When the same number of holes move forward at a faster velocity, the larger the amplitude of the individual waves, the larger their addition, and the larger the value of the amplitude of the collective wave: increasing (P).

When one object strikes another, part or all of the waves are transferred from one object to another. It is this transfer that creates the Conservation of Momentum.

When two objects collide, the transference of some of the waves is reflected back to the holes of the original moving object, and the direction of the original object is changed.

It is the wave within the surface of each individual hole the atom is made out of that is the real explanation of Conservation of Momentum. The amount of momentum transferred to another object is a function of the percentage of the collective sum of these individual waves that are transmitted to other objects during collisions.

#9 The Explanation of The Conservation of Angular Momentum

One of the world's most enduring mysteries is finally explained.

The explanation for the conservation of angular momentum has been waiting for a long time. Although our ancient ancestors did not know its technical name, they were able to observe its effects. Every time a person laid down upon the ice of a frozen lake, grasped the outstretched arm of a friend, allowed themselves to be spun around and then quickly pulled in their arms and legs, the conservation of angular momentum went to work, and they began to spin rapidly. This pirouette motion is most spectacular when performed by a highly skilled ice skater.

Although this fundamental law of nature could be expressed with mathematical equations, there was still no explanation of how it worked until now. And it is a most fascinating explanation:

When a three dimensional hole such as a proton moves through space, the internal wave upon its three dimensional surface [and the fourth dimensional waves within the fourth dimensional holes] moves it in a straight line.

However, when this distorted hole is part of say a ball attached to a string held in the hand of a man who is whirling it around his head, the hole is now moving in a circular path. The wave within the hole wants to continue to move in a straight line, and if the string broke, the hole would again continue to move in a straight line starting from the point of the break. However, because the string is strong, the ball moves at a constant speed, and the wave is reconfiguring at a constant rate. But if the man (while continuing to whirl the ball on the string around his head with his right hand) holds the other end of the string in his left hand and slowly starts to pull upon it, the length of the string between the man's right hand and the ball begins to shorten. This shortening of the string shortens the ball's radius of rotation.

As the string pulls upon the ball and the radius of rotation gets shorter, the pull on the string creates an acceleration that is transferred to all of the holes within the ball including the hole or proton we are observing. This acceleration creates an *additional distortion within the hole*, adding to the distortion that is already there, amplifying the size of the internal wave, making its space reconfigure faster, making it move faster.

The addition of these two distortions can be seen in the following drawings: Figure 9.1 shows the distortion if the hole was moving through space in a straight line; Figure 9.2 shows the added distortion created by the pull from the string; Figure 9.3 shows the vector addition of the two distortions; and Figure 9.4 shows the resultant vector distortion.

Figure 9.1

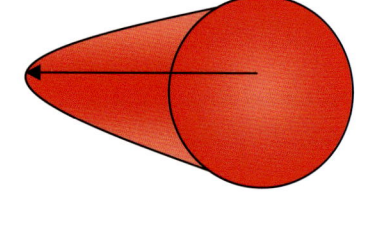

Figure 9.2

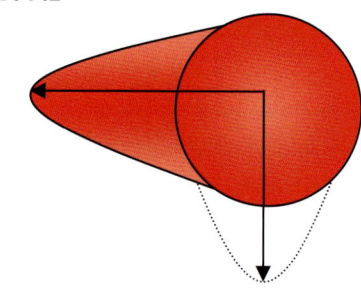

Figure 9.3

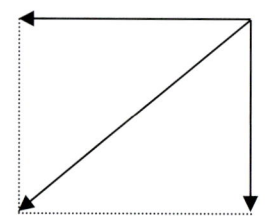

Figure 9.4

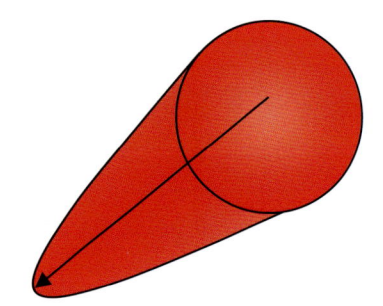

Because an acceleration has increased the amplitude of the internal wave within the three and fourth dimensional holes, the ball is now moving faster. As the string continues to shorten, the larger amplitude of the increasing wave continues to make it move even faster.

But so far, we have only been describing what is happening within the ball.

As the man pulls upon the string and the string pulls upon the ball, the ball pulls back upon the man through the string. Because the pull of the ball also distorts the holes within the man, his motion is slightly accelerated too. However, because the ball possesses a small number of holes in relation to the holes within the man, the distortion it transfers to the holes within the man is a lot less than the distortion transferred to the ball through the man.

Nevertheless, the distortion has been made within the holes within the man, changing his rotational speed, slightly accelerating him too.

But when the reverse happens, the distortions in the holes in both the man and the ball are ameliorated.

When the reverse happens, and the man loosens his grip on the string, allowing the ball to move away from him, the distortions are reversed.

Assuming a perfect transference of distortions [where none is lost to friction and no energy is dissipated as heat], as the man begins to let the ball move away from him, the holes it is made out of begin to be slightly distorted outward, in the opposite direction from whence they were originally distorted. This outward distortion negates their original distortion, and the ball begins to move slower. It is the same for the man.

When he began to loosen his grip upon the string and allow the ball to move away from him, its pull upon him lessoned as well. This lessoning of the pull decreased the distortions created within his holes, decreased their acceleration, and slowed his rotational speed.

Although the above illustration dealt with an example of a ball whirled above the head of a man, it must be noted that the exact same situation will occur with the holes of two objects held together by gravity [such as the Earth and the Sun].

Consequently, it is now easy to see how the increase and decrease in the distortions in the holes of two different objects connected together – increase or decrease the amplitude of their internal waves, allowing their speeds to accelerate or decelerate – creating the law called the Conservation of Angular Momentum.

#10 The Explanation of The Conservation of Charge

> One of the great mysteries of the subatomic world is finally and easily explained.

The discovery of the two vortices of space flowing back and forth between the electron and the proton explains many observations that were unexplainable before. The most obvious of these is the "conservation of charge".

The term "conservation of charge" is based upon the scientific observation that the break-up of a particle, or the change in a particle from one type to another [such as the decay of the neutron into a proton, electron, and an anti-neutrino] always leaves the same net charge. For example, since the net charge upon the neutron was zero, when the neutron was hit by the neutrino [see Figure 8.13], and changed into a proton and an electron, the respective charges of these two particles are positive and negative (the anti-neutrino has no charge). Hence, when a +1 and a –1 are added, they cancel each other out [+1 + (– 1) = 0]: making their net charge equal to zero.

For years, this *conservation of charge* was a mathematical observation in search of an explanation. But not any longer! The failure to explain this phenomenon resulted from a lack of knowledge about the existence of the vortex. But now, it can clearly be seen that these two "particles", (holes in space), are really the two ends of a swirling higher dimensional vortex constructed out of three dimensional space. Consequently, no matter how many times the vortex is broken, the number of additional entrances and exits into and out of three dimensional space will always cancel each other out, and their sum will equal the number originally present.

Hence it is easy to see how charge is conserved, and how another great mystery of science is easily explained.

#11 The Explanation of The Conservation of Mass and Energy

> Perhaps the most famous equation in the world is $E = mc^2$. We found out how it works when we built the atomic bomb. But we did not know *why* it works until now.

The explanation of the conservation of mass and energy is one of the longest and most awaited explanations in all of science. Luckily it is as easily explainable as the equation $E = mc^2$.

Mass is equal to the hole that is left when a volume of three dimensional space is removed to create the three dimensional hole; and *energy* is equal to the volume of the removed space.

Or, using ice cream as an example, taking a scoop out of the surface of a tub of ice cream leaves a hole in the surface: this hole is the equivalent of mass; while the volume of ice cream in the scoop is the equivalent of energy.

#12 The Explanation of The Great Mystery of Entropy

> Perhaps the single most important principle in all of thermodynamics is the great mystery: or, why does energy leave a high energy region and travel to a low energy region?

Entropy is the great mystery of Thermodynamics.

Like Newton's three Laws of motion, Entropy was another observation without an explanation. But it is a most significant observation. Entropy is the amount of energy that a system has used and will never be able to use again, such as the energy pouring out of the Sun.

Entropy is perhaps the single most important observation of energy ever made by science. It is the subject of books, lectures, college courses, and PHD Thesis and yet nobody knows why it is occurring! Nobody knows why energy is pouring out of the Sun and all of the stars in the sky. Nobody knows why heat is flowing out of all of the living creatures of the Earth; or for that matter, why all of the physical matter in the universe appears to be cooling off.

This massive phenomenon that is occurring everywhere in the entire universe is not only important to science, it is essential to life on earth. Because without entropy, the Sun would not shine, and we would not be alive!

[The explanation of Entropy was always very important to me because it was one of the first subjects I remember being told to accept without challenge because it was an Axiom: just another observation of the workings of the universe without an explanation. In fact, it was the first great mystery that made me realize a great mistake existed somewhere in science.]

Luckily things have now changed. With the discovery of the Vortex Theory, we do not have to blindly accept the mysteries of science – we can now explain them, all of them!

Although Entropy is a much discussed, "great mystery of science," according to this new Vortex Theory of matter it is reduced to a trivial subject.

According to the Vortex Theory, *Entropy is the result of the vortices trying to maintain their normal lengths; and the region of denser space surrounding the photon* [Figure 9.2].

Stated simply, when an atom is "struck" by a photon [Figure 9.1], the photon tries to add to the length of the vortices, finds it cannot, and is discharged back into the surrounding space.

Even if the dense space of the photon is retained within the vortex, elongating it, putting the electron into a higher orbital and thereby creating a higher energy state; the vortex collapses, and the extra volume of space is usually discharged back into three dimensional space within something like a hundred millionth of a second. Hence, atoms tend to "repel" rather than "collect" energy. Because of this fact, the majority of the photons within any system will eventually be thrown outwards into space, away from the matter they inhabit.

Also, the denser regions of space surrounding infra-red photons [heat], pushes them all away from each other, making them seek regions of space that have less photons [cooler regions].

#13 The Explanation of Dark "Matter

The explanation of Dark Matter explains the mysterious motions of Galaxy clusters, stars within galaxies, the orbit of the planet Mercury, and the orbit of satellites around the Earth.

DARK MATTER…

In the 1930's, astronomers Fritz Zwicky and Jan Oort, both working independently of each other, suggested that much of the matter of the universe could not be seen. The existence of this unseen or "Dark Matter" was based upon their observations of the motions of galaxies within gravitationally bound clusters of galaxies, and the motions of stars within the disk of our Milky Way Galaxy.

Because these stars and galaxies appeared to be moving much faster than was expected, they believed the gravitational effects of unseen, or dark matter were causing these motions. However, because the amount of dark matter appears to exceed the amount of visible matter by an amount of ten to one, this means there must be an incredible amount of unseen matter.

Although there probably is an enormous amount of matter orbiting stars in the form of planets, and most likely there are an enormous amount of burnt out stars, the ratio of ten to one is an incredible amount. If there were a vast population of burnt out stars in our galaxy, there should also be an incredible amount of white dwarf stars. Since we do not see a vast collection of white dwarf stars, all of these burnt out stars probably do not exist either.

Another explanation for this "Dark Matter" of the universe is **WIMPS** [**W**eakly **I**nteracting **M**assive **P**articles]. According to present day beliefs, the Milky Way Galaxy as well as every other galaxy in the universe is literally crammed full of these unseen particles. They are supposedly crammed in around us!

If this were true, if galaxies were indeed crammed full of these unseen "particles" as some present day scientists believe, their additional force of gravity could account for the faster motions of stars and galaxies, but they would create another problem – they would cause too much friction. Newton's first law of motion would not work. The hulls of spacecraft would heat up. The Mariner Spacecraft would never have made it out of the solar system.

Although such particles might possibly exist, their population probably doesn't.

The problem with WIMPS is that they are a product of the reasoning process tied to "Particle Logic". The vision of the universe where everything is made out of "particles". A system of mistaken logic that developed in response to Albert Einstein's incorrect belief that space was made of nothing.

The belief that everything in the universe is made out of particles seems appropriate for the subatomic world. But when we leave the subatomic world and travel into the world of stars, planets, and galaxies, the particle vision of the universe falls apart.

In the Galactic World of stars, planets, and galaxies the rotations of massive volumes of less dense regions of space easily explains what particles cannot. These rotations explain many of the universe's most unusual motions, especially those attributed to Dark Matter.

On the Galactic scale, the addition of all the massive volumes of less dense space surrounding all of the stars of a galaxy overlap each other creating a massive region of space to rotate as one object: like a record rotating on a turn table. All matter caught within the massive disk of less dense space rotates with it. This massive rotating disk of less dense space creates the effect that is attributed to "Dark Matter!"

Because these less dense regions of space also extend far outward into the space beyond a galaxy, clusters of galaxies behave the same way. Galaxies caught within the giant volume of rotating less dense space between other galaxies, rotate with it.

MASSIVE ROTATING DISK OF LESS DENSE SPACE EXPLAIN MANY OTHER STRANGE ROTATIONS…

There are a number of strange rotations of matter in the universe that cannot be explained by traditional Newtonian Gravity. Because the massive regions of less dense space that rotate with galaxies are created by the stars that make up the galaxies, these regions also exist around stars and planets. The less dense region of space closest to the Sun is responsible for the Precession of the perihelion of Mercury. While the less dense region of space closest to the earth is responsible for the slight movement of earthly satellites in polar orbits.

Consequently, the seemingly strange motions of stars, planets, and galaxies can now be explained by the motions of the space they are imbedded within.

When a star or planet moves through space, space is constantly reconfiguring around it, creating a spherical region of less dense space around it, making it appear as if this region – seen by us as its "gravitational field" – is traveling with it. However, it must also be understood that regions of less dense space exist, can exist *within* larger regions of less dense space!

For example, the Sun's "force of gravity" and the Earth's "force of gravity" are actually two individual regions of less dense space. The smaller region of less dense space surrounding the Earth exists within the much larger region of less dense space surrounding the Sun. Hence, one region exists within the other region. Consequently, if the larger of the two regions is itself in motion, it will affect the motion of the smaller region. Note, remembering that every proton, electron, and neutron making up a star or planet is a hole in space, it is easy to see how the motion of these holes are affected by the motion of the space they are imbedded in.

Applying these principles to the three problems presented above, allows us to explain the Precession of the perihelion of Mercury…

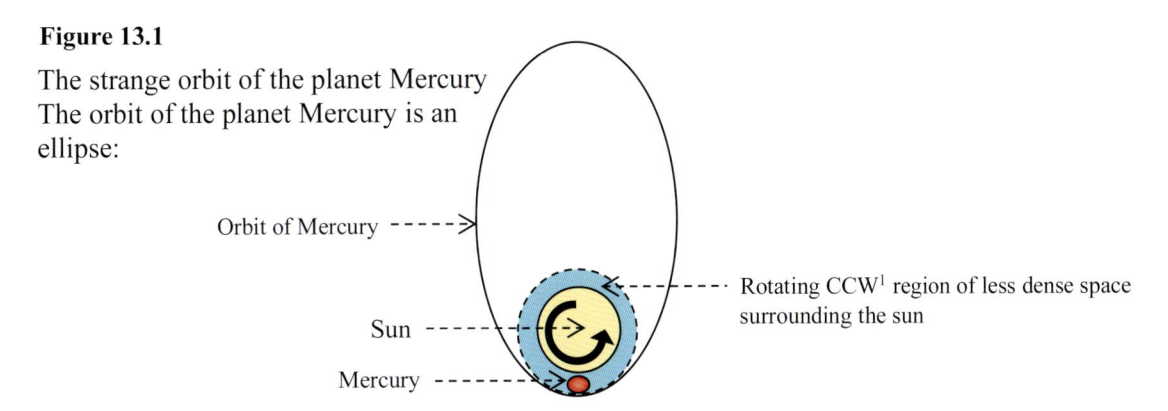

Figure 13.1

The strange orbit of the planet Mercury
The orbit of the planet Mercury is an ellipse:

Orbit of Mercury

Rotating CCW[1] region of less dense space surrounding the sun

Sun

Mercury

As Mercury orbits the Sun, the perigee of the ellipse progresses. However, this progression occurs at a much larger rate than Newtonian Gravity can explain. Therefore, Albert Einstein used the Theory of Relativity to explain this larger progression and the scientific community was satisfied. But with the discovery that time does not exist, that there is no 4th dimension called "Space-Time", the Theory of Relativity is now obsolete. Hence, there has to be another explanation for the strange rotation of Mercury.

The rotation of the Sun itself is the key to the strange rotation of the planet Mercury.

Directly around the Sun, a small region of the less dense space is in rotation. The rotation of the less dense region is created by the *diameter* of the Sun.

If an imaginary line could be drawn through the center of the Sun and all of the matter that touched this line could be seen, there would be a long thin string of matter about 850,000 miles long. As the Sun rotates, even though the reconfiguration of three dimensional space around each proton, electron, and neutron takes place at the speed of light, there is a time delay before the reconfigurations of space at either end of the string catch up to each other. Consequently, the region of space directly surrounding the Sun is constantly becoming denser as the Sun rotates slightly and then less dense as the "wave" of less dense space from the opposite side of the Sun passes through the center of the Sun and reaches the other side.

It is this time delay in the reconfiguration of space that causes the less dense space directly surrounding the Sun to be in a constant state of rotation. When the planet Mercury orbits the Sun, it is suddenly passing through space that is in motion, causing Mercury to move slightly CCW as it nears its perigee, causing its perihelion to progress.

---

[1] Where CCW refers to counterclockwise rotation.

Satellites orbiting the Earth in North to South polar orbits are affected by the rotation of the Earth's less dense region of space just as the planet Mercury is affected by the rotation of the Sun's less dense region of space.

Figure 13.2 Satellites Orbiting the Earth in polar orbits

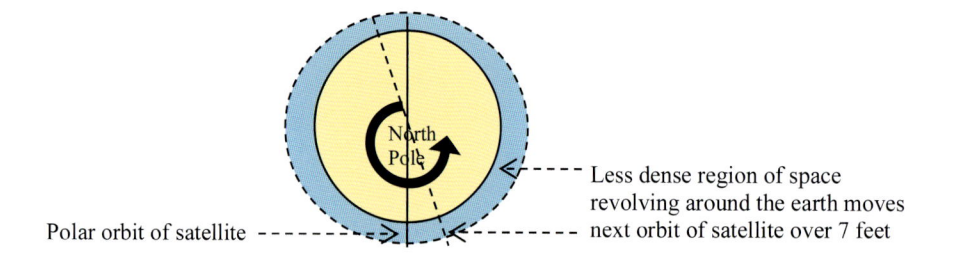

Polar orbit of satellite

Less dense region of space revolving around the earth moves next orbit of satellite over 7 feet

#14 Dark Energy

In the late 1990's, a picture taken from the Hubble telescope seemed to indicate that a supernova in a Galaxy located in a distant part of the universe was accelerating away from us at a much faster rate than it should. But how can this be? How can all of the previous spectrographic analysis of the pictures from the last 50 years showing the Red Shift of the Galaxies indicate one result, and this picture indicate another?

The answer to this dilemma can be resolved when it is realized that this picture was taken on the other side of a region in the universe where no galaxies can be seen with earth bound telescopes. This indicates that there is a vast sea of space between the Earth and this distant Galaxy.

In these vast reaches of space where there are few Galaxies, the space is *denser*. Hence, it tends to cause all of the matter in the Galaxies located on all sides of it to accelerate away from this vast wasteland towards *less dense* regions of space.

Figure 14.1

Galaxies

As will be explained in Discovery #30, because the space of the universe continues to expand, vast regions of empty space grow larger, become denser and increase the expansion rate of the matter on either side of it: increasing the expansion rate of the entire universe.

It can now be seen that it is the increase in the density of the space of the universe that is responsible for creating the increase in the expansion rate of the universe; and not the present mistaken belief in something called "Dark Energy". Stated simply, dark energy doesn't exist. The belief in Dark Energy is another false concept created by the erroneous logic of 20th Century science.

#15 Anti-Gravity Engineering

> The understanding of this force not only explains many unexplained phenomena of the universe such as those found in Chapter 15, but it will also finally allow us to travel to the stars!

Because the anti-gravity force was considered to be too important of a principle to wait until the end of this book, it was presented in Chapter 15. However, because the discovery of the anti-gravity force is a historical discovery, and because of its looming importance to the future of all mankind, the following is written to inspire the next generation of scientists and engineers.

For the scientists and engineers of the future:

Just like the discovery of the knowledge that revealed we were living at the bottom of an ocean of air led to the design and the development of the airplane wing, the discovery that space is made of something, and gravity is created by less dense and flowing space will lead to the development of anti-gravity lifting and propulsion devices. Devices that will allow a housewife to pick up a car, and spacecraft to "FALL" off of the surface of the Earth into orbit!

Just as gravity causes objects to fall to Earth, anti-gravity technology will make objects fall upwards off of the earth and into outer space, and then accelerate to near light velocities! This anti-gravity technology will inspire an industrial revolution that will eclipse previous industrial revolutions. This new industrial revolution will renovate and revolutionize the entire world. The future it will create is almost unimaginable!

Just as the electronic technological advancement of our era is almost unimaginable to the generation that fought the Civil War, the future will be nearly as unimaginable to the generation that fought the Viet-Nam War.

If you told soldiers standing upon a Civil War battlefield that one day, craft larger and heavier than train locomotives would fly through the air at speeds faster than the speed of sound, and that the horse and buggy will become obsolete, these men would think you were crazy.

But changes to man never seem to bring changes to men. If you told soldiers standing in a rice paddy during the Viet-Nam War that within a hundred years, the car, the plane, and most of the heavy machinery of the 20th Century will become obsolete, they would probably think you had gone mad. Or, you were trying to get a medical discharge out of the Army. Nevertheless, these inevitable changes will occur.

Anti-gravity technology will revolutionize all modes of transportation. Supertankers and Aircraft Carriers will fly above the oceans. No longer will they have to sail through the Panama Canal; they will sail above it! They will cross over land as easily as water. Reefs and shoals will no longer be hazards to navigation.

Most shocking of all will be the fact that anti-gravity technologies will not only make cars and planes obsolete - but concrete roads and freeways will no longer be needed either!

Arrays of antennas generating anti-gravity bends in space will protect the Florida coastline from raging hurricanes. Similar arrays on boats and upon satellites in space will "herd" rain-laden clouds to drought areas. No longer will crops die or people starve!

Space flight will become as common as airplane flight is today. However, unlike the problems experienced today, future space flight will be much different than the rocket propelled technology experienced by present day astronauts. Anti-gravity technology will allow constant acceleration over the entire length of the trip. This constant acceleration will end the weightlessness experienced by the Astronauts of today. Spacecraft will take off slowly and land slowly. Anybody will be able to fly into space.

Constant acceleration will make speeds that are impossible today become commonplace in the future. If somehow you could go back in time and tell Magellan during his record three year around the world trip – that someday men will circumnavigate the globe in a little more than a day – there is no way he would believe you. And yet now we see it done every day.

Likewise, antigravity technology will allow us to accelerate to near light velocities. These speeds will allow us to colonize Mars, our Moon, and the Moons of Saturn and Jupiter. One day, or rather, "one night", the people of the Earth will be able to look up into the night sky and see the lights from the cities upon the Moon!

To hell with West Palm Beach, the place to have a condo will be on the North Pole of the Moon or upon the mountains of Mars! Some wealthy eccentrics will probably build haciendas upon the distant moons of Jupiter and Saturn. No longer will they have to have locks on doors or security alarms. Others, like Columbus of old, will embark upon fantastic voyages of discovery. They will travel at near light velocities to the planets orbiting distant stars. Perhaps two such people will be another world's Adam and Eve.

Perhaps it will be you! If you are a teenager in school planning to go to College soon, or if you are a student in College now, think about the future you want to create. Be a champion. Dare to be great. Do not let the limitations of others drag you down.

When you study science and engineering think about how to apply what you learn to creating artificial anti-gravity, and you may be the one who discovers it!

#16 What the Neutrino Really is and Why it Possesses both Matter and Energy Characteristics

The mysterious phantom particle of nature that Wolfgang Pauli named the *Neutrino*: ("The Little Neutral One"), is a mystery no more.

The neutrino has always been intriguing to science because it possesses both matter and energy characteristics. Because it travels at the speed of light, it looks like energy; but because an anti-neutrino can strike a proton, creating a neutrino and a positron, or because a neutrino can strike a neutron creating a proton and an electron, the neutrino seems to behave like matter. However, the answer is cleared up when it is realized that the neutrino is really a longitudinal wave.

Just as a photon can be compared to a compression wave traveling through the interior of a body of water, a neutrino can be compared to a longitudinal wave traveling along the surface of the water. But in this circumstance, the wave is traveling upon the surface between three dimensional space and higher dimensional space.

A neutrino is created by the deflation of a volume of space such as when a neutron "deflates". The word "deflates" was used because when a neutron decays into the proton, the electron, and an anti-neutrino; some of the larger internal volume of the neutron deflated, causing a depression in fourth dimensional space. This depression is the anti-neutrino.

Because the neutrino is a "quanta-sized" wave, it is not radiating outward in all directions at once like a surface wave created when a rock is dropped into the ocean. Hence, for lack of a better analogy, neutrinos can be said to be a "quanta-sized splash" upon the surface of higher dimensional space.

The two dimensional to three dimensional representation of the neutrino is drawn below. It can be compared to half a sphere bent upwards into three dimensional space, while an anti-neutrino could be compared to a half sphere bent downwards into three dimensional space. Neutrinos possess ½ spin because they possess a fourth dimensional axis of rotation [see discovery #4].

Figure 16.1

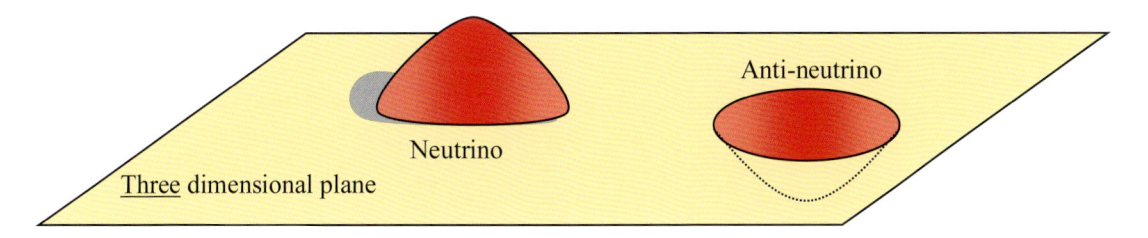

The reason why there are three types of neutrinos comes from the fact that the electron, the Muon and the Tau leptons create bends in higher dimensions of space. This will be explained in Book 3, *The Quark Theory*.

#17 The Explanation of Buoyancy

2200 years ago, the Ancient Greek Mathematician and inventor Archimedes discovered the phenomenon of buoyancy; and in his excitement, supposedly ran down the street naked yelling "Eureka"; but neither he, nor anybody else knew how it worked until now.

We all know about the phenomenon called Buoyancy, but until now, no one knew why this phenomenon existed. Although it was touched on before in this book, because it is so important it is listed again in these discoveries.

Buoyancy is important to us all. Without buoyancy, no ship could sail the sea, nor could anyone pan gold. Buoyancy is another one of those phenomena of nature that is so old, its acceptance is without question. We know that it exists, and because it is a phenomenon that is so common, we don't even think to question it.

But why does buoyancy exist? What is the mechanism in nature that is responsible for making an object "Buoyant"? What makes something float upon the surface of the water? What makes minerals of different densities separate apart from one another in a revolving fluid?

The answer is something nobody has ever suspected before.

It is not the mass of the object (such as a piece of wood) that causes buoyancy, but the density of the space within the wood! It works like this: the *more* protons and neutrons per cubic inch, the *less dense* the space within; and the *fewer* protons and neutrons per cubic inch, the *denser* the space within. Therefore, it can be said that within a cubic inch of for instance a mineral containing a lesser number of protons and neutrons than another cubic inch of a different mineral containing more protons and neutrons, regions of denser or less dense space are in effect "trapped."

This relationship can be seen in Figures A and B:

Figure 17.1

FIGURE A   FIGURE B

o o o o o o   o   o   o   o
o o o o o o   o   o   o   o
o o o o o o   o   o   o   o
o o o o o o   o   o   o   o

In Figure A, more matter per cubic inch creates a region of less dense space in comparison to Figure B. In Figure B, less matter per cubic inch creates a region of more dense space in comparison to Figure A. Within a stream of flowing water, when these different regions of space are being mixed together, or allowed to rotate together, the turbulence of the water allows them to rearrange their locations. The more dense regions of space found in less massive rocks try to move upward, while the less dense regions of space trapped in more massive rocks move downwards. In effect, *when moving*, these denser regions of space possess anti-gravity properties.

It must be emphasized that even though all of the protons and neutrons from both cubic regions of space are accelerated towards the Earth's center of mass equally, when mixed together, it is the denser region of space that is bent upwards and away from the direction of the Earth's center of mass. Consequently, this denser region of space that is trapped within a less dense rock seeks to move upwards and away from the Earth creating a buoyancy effect.

The same effect is produced within the hull of a ship. A region of more dense space is "trapped" within the hull making it push up out of the water.

A hot air balloon rises because photons are dense regions of space. When a massive amount of them are trapped within an enclosure, such as a balloon, the region of air that is constantly exchanging them has in effect entrapped a volume of denser space.

#18 The Explanation of the Ionic and Covalent Bonds in Chemistry

> The now easily understood explanations of the bonds that hold matter to matter are not only fascinating and extremely important information for every chemist, but also, for everyone who loves to see a great mystery solved.

Within an atom, the position of electrons in relationship to the nucleus is determined by the following factors:

1. The denser space surrounding the electron.
    a. The denser space surrounding the electron creates the repulsive force between electrons.
    b. The denser space surrounding each electron creates a barrier between it and other electrons.

This region of denser space acts as a volume that only allows so many electrons to occupy a shell. Once the volume of a shell is filled up with the volumes of the denser spaces surrounding the electrons in the shell, no more electrons can fit into the shell. What is important to realize is the fact that even though the electron is a tiny object, the denser region of space surrounding the electron makes it occupy a much larger volume.

2. The length and position of the vortices.
    a. The length of the vortex connecting an electron to a proton restricts its motion. It can only move so far away from the proton before the volume of the space flowing into and out of the electron has a significant enough change to draw it back towards the nucleus.
    b. The positions of the three dimensional vortices flowing from electrons to the protons flows past shells, restricting the motions of the electrons within the shells.

3. The motion and position of other electrons.
    a. The motions of nearby electrons changing their position, changes the position of the regions of less dense space surrounding them, and pushes other nearby electrons away from them, changing their positions as well.

4. The position of the proton in the nucleus that the electron is connected to.

Losing and gaining electrons.

When an atom loses an electron, the proton [hole] it was connected to in the nucleus is no longer neutralized and begins to pull space into it. The flow into the proton decreases the volume of the space immediately surrounding it – the same space the atom is constructed out of – decreasing the size of the atom.

This change in size is seen when lithium (Li) loses its single valance electron turning it into Li^+. When this situation occurs, the lithium atom decreases to less than one third of its original size. It should be noted that even though the electron is no longer attached to the atom, the proton within the nucleus that the electron was connected to is still connected to this lost electron via its higher dimensional vortex.

When the atom gains one, two, or three electrons to complete its valence shell, a different situation occurs – the atom enlarges. For example, when oxygen (O) gains two valence electrons turning it into O^{--} its diameter swells to more than three times as large as when it was electrically neutral.

The atom enlarges because of the space flowing out of the two added valence electrons. The space flowing out of these electrons subtracts from the less dense space of the atom allowing the more dense space from the other electrons of the atom to enlarge the size of their shells – enlarging the atom.

The act of one atom pulling electrons off of another atom, creating ions out of both and then causing them to be drawn to each other is referred to as *electrovalent bonding* or *ionic bonding*. In the past this curious behavior has been attributed to what was called an atom's "electro-negativity." It was believed that non-metals possessed more "electro-negativity" than metals. Consequently, because a non-metal atom such as chlorine (Cl) would tend to be more electro-negative than the metal atom sodium (Na), when the two atoms came together under the right circumstances, the difference in the electro-negativity would cause the sodium to lose its one valence electron to the chlorine's seven valence electrons completing its valence shell of eight electrons. This situation would turn the sodium into a positive ion Na$^+$, and the chlorine into a negative ion Cl$^-$. Because these two atoms now possess opposite charges, the electrostatic force of nature would pull them together, binding them together forming sodium chloride Na$^+$Cl$^-$ commonly called table salt. However, using the reasoning processes of the Vortex Theory, it is now easy to see that this entire process is the result of what can now be called a *"Gravity Well."*

A *Gravity Well* is created when an atom's valence shell is one, two, or three electrons short of the eight electrons needed to fill it. These "vacancies" cause the less dense space of the atom to have voids upon its surface. These voids are intense regions of less dense space. These intense regions are in effect holes or wells of less dense space – hence, "Gravity Wells."

Upon the surface of an atom that possesses only one, two, or three electrons in its valence shell, the opposite condition arises. In this situation, the less dense space surrounding the atom possesses either one, two, or three "Gravity Spikes" created by these electrons. However, even though the more dense space surrounding electrons creates an anti-gravity effect, because the electrons are really holes in space, the surface of the electrons are distorted in the direction of the gravity wells. This distortion of the electrons is greater than the distortion created by the weaker bond that holds them to the surface of their atom and they move to the surface of the other atom.

When the electrons switch atoms, they create the positive and negative ions that pull the two atoms together, binding them together. It should also be mentioned that even though the electrons have switched atoms, they are still connected to the protons in the first atom via their higher dimensional vortices.

A similar situation occurs when a *covalent bond* is formed. Covalent bonds are created when gravity wells of two different atoms pull the atoms together. These intense regions of less dense space are found mainly in nonmetals and produce molecules such as Cl_2 O_2 N_2 SO_2 CO_2 CO SO_3 or CCL_4 etc. In each of these molecules, it is the gravity wells created upon their surfaces that pull one atom into the other, completing their valence shells – making it appear as if they are somehow sharing electrons as is commonly, and mistakenly believed.

#19 The Mathematical Explanation of The Michelson Morley Experiment

The mathematical explanation that took eight years of work and is the proof of the Vortex Theory was awarded a PhD in 2005 by the Russian Ministry of Education.

The explanation of the Michelson Morley experiment is found in the proof at the end of this book. Although the mathematics of this proof seems simple, it took almost eight years of work to construct this proof.

Because Albert Einstein's explanation of this experiment is a mistake, the real explanation is one of the most significant discoveries ever made in science. Hence, it deserves mention here in this section on discoveries.

#20 The Explanation of: The Strong Force, The Weak Force, The Electromagnetic Force, The Force of Gravity and The Anti-Gravity Force

> For the first time since they were discovered, the four forces of nature are not only explained, but they can also now be drawn! How the four forces of nature are created is one of the most beautiful visions in the subatomic world.

Again, just as with the explanations of the proton, electron, and the neutron, and the Michelson Morley experiment, the forces of nature need to be in this section.

Although it was necessary to already explain the forces of nature in previous chapters, they have been placed in this section because their explanation is another one of the greatest discoveries ever made.

The Force of Gravity: Chapter 11

The Electromagnetic Force: Chapter 12

The Weak Force: Chapter 13

The Strong Force: Chapter 14

The Anti-gravity Force: Chapter 15

#21 The Explanation of Mass

> Just as previously mentioned, regarding the proton, electron, and neutron, the strong force, the weak force, the force of gravity and the electromagnetic force, the Michelson Morley experiment and buoyancy, the explanation of mass is also one of the most important discoveries of the Vortex Theory. Consequently, even though it has already been explained in Chapter 16, it is listed here because it has never been explained before. Note: there is no Higgs Boson Particle!

#22 The Explanation of Energy

> Again, even though energy is a concept used throughout the entire world, nobody knew what it was until now. Although it was already explained in Chapter 9, it is also listed here with these other blockbuster discoveries of science.

#23 What causes acceleration

> Not even the great Albert Einstein who equated acceleration with the force of gravity knew *how* acceleration was created: but now it is easy.

One of the greatest discoveries of the Vortex Theory was acceleration. Like many of the other great discoveries made by the Vortex Theory, it has already been explained [Chapter 11]. It is listed here because of its great importance.

#24 The Explanation of the Muon's Prolonged Lifetime when moving at relativistic velocities

> The prolonged lifetime of the Muon is listed here instead of Book 3, because many scientists said they would like to see its explanation rather than wait for Book 3.

The lifetime of the Muon is .000002 seconds before it decays into an electron, a neutrino and an anti-neutrino. However, because a Muon is a hole in space, when this hole moves through space, space reconfigures around it.

This reconfiguration takes place at the speed of light. However, at extremely high velocities approaching the speed of light, the vector flows tangent to the particle can no longer reconfigure at the speed of light. To do so would make the resultant vector exceed the speed of light, exceeding the speed of light would exceed the elastic modulus of space, creating a rip or "tear" in space.

Consequently, when space flows into a hole, out of a hole, or reconfigures around a hole, it has to do so at the value of the "y" vector. Because this is a slower reconfiguration speed, the decay of the Muon also slows: [in the figures below, notice how in Figure 24.1, how space can flow into the Muon at the speed of light (C). But now, when moving, notice how in Figure 24.2, space can only flow into the Muon at the speed of the Y vector.]

Figure 24.1   **Figure 24.2**

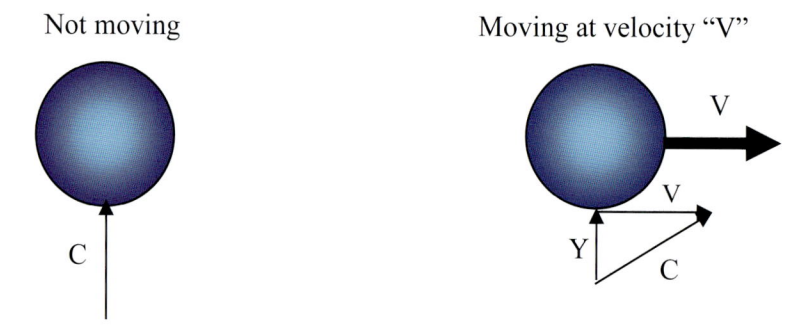

Not moving   Moving at velocity "V"

This value of the "Y" vector is equal to: $Y = \sqrt{1 - V^2/C^2}$ [Where "C" equals the speed of light, and "V" is a % of C.]

Because the slower time that it takes to decay is exactly equal to the value of the time constant proposed by Lorentz and Einstein, it appears as if "time" is somehow responsible for this phenomenon. However, as can now be seen, this is not true. Instead, this phenomenon is created by the slower reconfiguration time of space.

#25 All of The Phenomenon Associated with The Theory of Relativity are Now Explained

> The twin paradox, the slowdown of clocks, the increase in mass when traveling at near light velocities and the unusual orbit of the planet Mercury are just some of the relativistic phenomena of the universe that the Vortex Theory easily explains.

#26 Why All Charged "Particles" Possess the Same Amount of Charge

> Never suspected by 20th Century science, the real reason why all sub-atomic particles possess the same amount of charge is a stunning triumph of the Vortex Theory.

One of the most perplexing phenomena in the subatomic world is the relationship between charge and particle size. It has long been a mystery why charged particles of all different sizes possess the same amount of charge.

Some believe that the discovery of quarks within protons, anti-protons, neutrons, and other barons has cleared up this mystery. It is currently believed that quarks have one-third and two-third charges. It also appears from the discovery of the Vortex Theory that an electron has the same charge as the proton because it is connected to the proton by the higher dimensional vortex.

But even though the proton and electron are connected by a higher dimensional vortex, the mystery is still not solved. For when two gamma rays collide, they create an electron - positron pair that has no connection to protons or quarks. Yet both possess the same charge that proton, anti-proton pairs have.

Furthermore, when the much larger and more massive Muon, Anti-Muon pair is created, or when the even larger and more massive Tau, Anti-Tau pair is created, both sets of particles possess the same charge as the much smaller electron - positron pair.

This observation leads us to conclude that charge is not the result of the size of the particle, the mass of the particle, or the quarks carried by a particle: instead, charge is the result of some other phenomenon of nature. Because charge is a function of flowing space, there must be some sort of relationship between the size of the particle and space's ability to flow.

And by using the concepts of less-dense and flowing space, we discover a wonderful trick of nature - beautiful in its simplicity. Here is how it works:

When a hole is made in three dimensional space, its creation is completely different from the holes we are used to creating in a material such as a piece of wood. Unlike a hole in a piece of wood, a hole in 3d space is not created by removing material, but by pulling 3d space inward or pushing it outward.

When a hole is opened up in 3d space, it begins as a "pinpoint" and expands outward. For a 3d hole that space is flowing into, as it expands, the surrounding space is pulled into the hole. This creates the less dense spherical region that is responsible for generating the "force" of gravity. Because the force of gravity is inversely proportional to the square of the distance, the density of this region is inversely proportional to the square of the distance away from the hole, forming a density gradient.

Far away from the hole, "normal space" is at its normal density. However, the closer one gets to the hole, the less dense the space is. And this change in density creates a change in the elasticity of space.

The relationship between the density of space and the elasticity of space is extremely important. The less dense the space is, the less elastic it is. And the less elastic it is, the slower it flows. The larger the hole the larger the region of less dense space surrounding it, and the slower the flow into it; the slower the velocity of the flow - the smaller the volume of space flowing into it. Consequently, the volume of space flowing into different sized holes is exactly the same!

It is also important to note that unlike other mediums such as water, the flow of space cannot accelerate into the hole as the hole is created. Because the elasticity of 3d space decreases, as the hole grows larger, its velocity actually decreases, making the space decelerate into the hole. Hence, the flowing space cannot create a larger and larger hole!

For space flowing out of a hole, a different situation occurs.

The creation of a hole that space flows out of also starts as a "pinpoint." However, in this instance, space is pushed outward, increasing its density. And just like the less dense space surrounding a hole that space is flowing into, a density gradient is created around a hole that space is flowing out of. This density gradient is directly proportional to the size of the hole. So, the larger the hole, the greater and larger the gradient!

At the beginning of the hole, the gradient is at its maxim value and then drops off as it approaches "normal density" space. This maximum density near the hole creates a new situation. Because space is denser, its elasticity is increased. This increase in elasticity allows space to move faster. This faster moving space crosses the gradient quicker, allowing it to arrive at the normal region of space beyond the denser region at the same time as if the space were flowing from out of the "pinpoint" in the exact center of the hole. This situation makes the flowing space arrive at the same distance from the hole at the exact same time no matter what the size of the hole. Making the measured value of the electrostatic charge the same for all "particles" of different sizes.

Note: when a proton captures an electron creating a hydrogen atom, the less dense region surrounding the proton, and the denser region surrounding the electron cancel each other out, creating a steady velocity of flow out of one hole and into the other.

#27 The Explanation of Black Holes

The discovery of what a black hole really is will shock the scientific world! Because protons are holes bent into space, and electrons are holes bent out of space, they cannot "join" to create one hole. Instead, as electrons and protons are crushed together within the interior of a supernova, they create a giant neutron. This giant neutron has mistakenly been called a black hole in space.

#28 The Explanation of Planck's Constant

> The great German researcher's constant can now be explained.

In the formula E = ℏv, ℏ = Planck's constant, E = energy, and v = frequency of light. This formula can also be expressed as ℏ = E / v. When the formula is changed, notice how Planck's constant is directly proportional to energy and inversely proportional to frequency. Since *h is a constant*, if the frequency of a photon increases, its "energy" [its dense volume of space] has to increase; and if the energy decreases, the frequency [ability to expand and contract] has to decrease.

Because the Vortex Theory reveals that energy, or rather photons are nothing more than "bunches" of very dense space surrounded by a massive region of dense space, Planck's constant is revealed to be the ratio between the amount of dense space that we call energy and its frequency of vibration [ability to expand and contract]! (See Chapter 9)

#29 The Explanation of High Velocities and Increasing Mass

> This fantastic phenomenon occurring at near light velocities now has an explanation!
>
> At extremely high velocities approaching the speed of light, the mass of a "particle" increases. However, the increase in mass is not being created the way science presently believes.
>
> The slower apparent velocity of the space reconfiguring around the hole ["particle"] as it is moving through space creates the increase in the mass of a particle that is moving at an extremely high velocity. The creation of this effect is most fascinating:

As a hole moves through space, the surrounding space is constantly reconfiguring around it. For example, as the proton moves through space, the surrounding space is first bent inward towards the proton as it approaches, and then back outward again, in the opposite direction as the proton passes by. These two motions are extremely important.

It is extremely important to realize that two different motions of space occur, with each motion directly opposite to the other. It is also important to understand that the reconfiguration of space around the hole is taking place at the fastest speed that three dimensional space can move - the speed of light: a situation that creates a problem when a hole is moving near the speed of light.

When a hole is moving at extremely high velocities, the surrounding space can no longer reconfigure at the speed of light. If it did, the vector addition of its speed, and the speed of the hole would produce a resultant vector whose scalar value would surpass the speed of light. Such a vector would exceed the vector representing the elastic modulus of space, creating a rip or "tear" in the surface of three dimensional space, creating more holes [particles].

Therefore, when a hole is moving at near light speeds, the three dimensional space surrounding it begins to reconfigure at a slower rate. This slower rate is a function of the two motions of space that are taking place: the motion of space towards the hole, and the subsequent motion of space away from the hole.

What is most fascinating is the fact the mathematics of the space flowing inwards towards the hole, and then back outwards away from the hole is exactly the same as the mathematics that describes the action of the space flowing into the proton and out the electron in a hydrogen atom. Hence, its mathematical analysis requires the exact same mathematical proof that is found in the appendix of Book 1 that describes the flow of space into the proton and out of the electron, also describes how the surrounding space is moving first towards, then away from the proton and the electron as they move through space.

For now, the increase in mass can be represented by the vector flow that is perpendicular to the motion of travel in the figure below.

Consequently, it can be said that when space flows into a hole, out of a hole, or reconfigures around a hole, it has to do so at the value of the "Y" vector [see the PROOF].

Figure 29.1   **Figure 29.2**

Not moving   Moving at velocity "V"

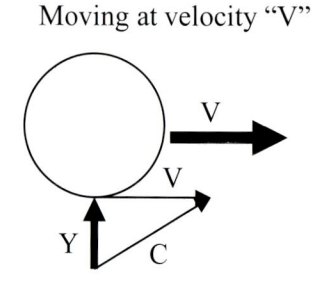

This value of the "Y" vector is equal to: $Y = \sqrt{1 - V^2/C^2}$ [Where "C" equals the speed of light, and "V" is a % of C.]

Because space has to both move and flow slower by the value of the Y vector, it reconfigures slower by the value of the y vector. Since it cannot bend and flex as fast as it could, its elasticity changes. Because it is now less elastic, it takes more "force" to bend it – increasing its opposition to being bent – increasing its "mass"!

#30 The Explanation of The Striking Parallel Between Newton's Law of Gravity and Coulomb's Law

> It is no coincidence that Newton's Gravity equation and Coulomb's Law of Charges are almost identical!

In the 18th Century, when Charles Coulomb discovered the law for the force of attraction and repulsion between two electric charges, nobody was probably more surprised than him to realize afterwards that it was nearly identical to Newton's universal law of gravity:

Newton's law of gravity: $F = G m_1 m_2 / r^2$; Coulomb's law of charges: $F = k_c q_1 q_2 / r^2$

The only difference between Newton and Coulomb's law are the constants G and k_c and the fact that the masses m_1 and m_2 are replaced with charges q_1 and q_2 while division by the distance between them squared (r^2) is the same for each. The reason for these two strikingly similar equations comes from the fact that they are both describing "forces" being created by the same set of circumstances but under different conditions.

In each case, the force F is being created by distorting the spherical holes into pear shaped holes. However, the distortion is happening under two different conditions. For the force of gravity, the distortion of the holes is being created by the regions of *less dense space surrounding the holes*; while for the force of electrostatic charges, the distortion is being created by the *space flowing into or out of these holes*.

The reason why both masses are multiplied together, and both charges are multiplied together, comes from the fact that both mass and charge represent spherical regions of dense and less dense space surrounding the holes in one object that are responsible for creating distortions in the holes of the opposite object.

The distance between the masses or the charges (r), divides both formulas because the spherical regions of dense or less dense space that surround the holes form a density gradient that decreases in intensity further away from the hole. The reason why the value of **r** is squared comes from the fact that the "force" exerted upon one object by the other object is in fact distorting the surface of the other object.

What a "force" is really doing is distorting the square area of the sides of the holes of the opposite object. Hence, the decreasing density gradient is a function of the increasing surface area of a sphere of radius **r**. As the sphere increases in size, its square area also increases as a function of r^2; but as the square area increases, the density of space increases by a factor of r^2: for the Force of Gravity, the region of less dense space surrounding the object becomes *more dense* with the increase in the square area decreasing its ability to distort the surfaces of the holes of the other object; while for the Electrostatic Force, the increasing square area decreases the gauss per square inch, decreasing its ability to attract or repulse the gauss of the other object – decreasing the ability to increase or decrease the flow of space into the opposite object, which in turn decreases the ability to distort the shapes of the holes in the other object.

Because both the force of gravity and the electromagnetic force distort the surfaces of three dimensional holes, both constants **G** and **k_c** contain as part of their value the unit of 4π [a value related to the circular sphere surrounding each force.]

#31 The Explanation of The Creation Of The Universe

Originally the creation of the universe was going to be put into this book. However, and shockingly so, the explanation of the creation of the universe is discovered in the creation of the particles out of which all matter is made. Some of this information is detailed, such as the charges upon quarks, the hierarchy of quarks, the number of dimensions that exist and why, and a few other things. Consequently, because all of this is explained in the second book, the explanation of the creation of the universe has to wait until this information is presented. However, this much can be said, the creation of the universe is unlike anything anybody ever imagined before!

#32 The Explanation of Time and Time Dilation Effects

> Contrary to present and historical belief, time is not a fundamental principle of the universe; instead it is a function of motion, a phenomenon created by motion. This was the pivotal discovery of the Vortex Theory that leads to all of the other discoveries listed in these books. However, *this explanation is found at the end of the Proof in Book 1*. It is listed here so that non-mathematical readers, who might avoid the Proof, will look through the Proof to find it and will not miss out.

#33 The Grand Unification Theory? I don't think so!

> It is amazing how errors are passed from one generation to another without either generation knowing what is happening. We all know about the earth being flat error, and about the earth being the center of the universe error. We smile at the ignorance of past generations because we now consider ourselves to be so smart. But are we? I'm afraid not.

When it comes to uniting the "Forces of Nature" we are found to be just as ignorant and lacking as our ancient ancestors.

This problem stems from the fact that there is no "Force" in the universe that the known forces are a part of.

What is identified as Force are just configurations of bent and flowing space that cause mass to be attracted to each other or repelled away from each other. For example…

The force of Gravity: it can now be seen that the force of gravity does not "pull" one mass towards another. Instead, it is less dense space that causes distortions in the holes that matter is made out of. These distortions accelerate one mass towards another, there is no pull.

Then comes the *Electromagnetic force*: and again, in this instance, it is flowing space that is the cause of distortions in the holes that matter is made out of. And again, it is not a "Pull" that attracts matter to matter, but the acceleration of matter towards matter that brings matter together.

Next is *The Weak Force*: here we see that a sharply bent vortex of flowing space breaks, recreating the proton and the electron.

And finally comes the *Strong Force*: in this instance, protons and neutrons are held tightly together in the nucleus of the atom by an amazing switch of identities between particles.

Also, we discover the Anti-Gravity force: but again, this force is created out of a region of dense space that accelerates matter away from other matter.

In all of the above cases, these forces of nature are not creations of one universal force, but rather, just configurations of bent and flowing space that seem to pull or push matter towards or away from each other. Hence, there is no unification!

Something else must be said too: one of the great mistakes of 20th Century science is the belief that the forces of nature are transferred by particles. This is a mistake. In fact, nothing could be further from the truth. The particles called Gage Bosons are the effects of "force" and not the cause of "force". Gauge Bosons are "by products" of force. They do not transfer force; this is another myth. The photon does not transfer the electromagnetic force; there is no graviton; there is no Higgs Boson particle; and the W and Z particles will be discussed in Book 3: The Vortex Theory.

#34 A Neat Trick of Nature! Why Matter Does Not Flow in The Miniature Rivers of Flowing Space!

> We all know that when we throw a stick in the river, it flows along with the river like it is a part of it. So, if electrostatic and magnetic charges are really flowing space, how come when matter is introduced into the flow, it does not flow too, like the stick? It doesn't, because of a neat trick of nature!

Long ago, when I first started upon this quest, and made the discovery that electrostatic and magnetic lines of force were really tiny flowing "rivers of space", I was bothered by the fact that if so, then why didn't matter flow along with these miniature rivers?

It took many years before I found the answer, then it was a conundrum, but now it all seems all too easy. Here is the explanation…

The key to understanding what is happening is to remember that all matter: protons, electrons, and neutrons are holes in space. Consequently, when a proton is part of the nucleus of an atom, [say a hydrogen atom for example], its electric charge is neutralized by the electron. Hence it can now be just considered to be a hole in space: a hole in space that space is constantly reconfiguring around as it moves.

So, when a powerful electrostatic charge is brought close to it, a strange and wonderful effect begins to occur. The flow of space in the direction of the atom wants to get around the hole. In the figure below, the circle below represents a cross section of a proton, the arrows, the electrostatic field:

Figure 34.1

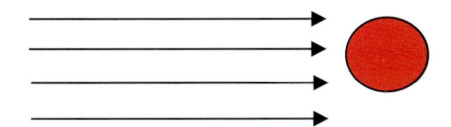

At first glance it appears as if the flowing space will push the proton off to the right of the page. However, this doesn't happen because the hole is a void in space. So instead of pushing the hole, the flowing space has to flow around it:

Figure 34.2

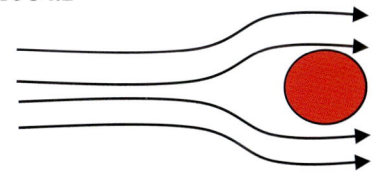

However, when it flows around the proton, it has to split. See the two ended arrow below:

Figure 34.3

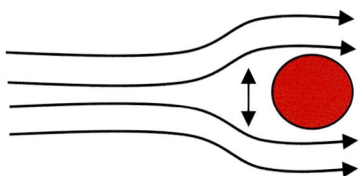

And when the space splits apart, the hole that is the proton is distorted in the direction of the split:

Figure 34.4

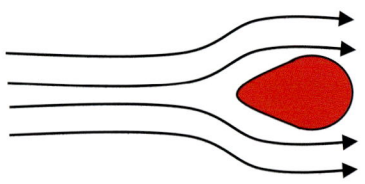

This distortion accelerates the proton in the direction of the flowing space. The greater the volume of the flow [greater the flux], the greater the distortion. Maintaining its position, keeping the proton from being swept away! A simply neat trick of nature!

This same situation happens with magnetic fields. Also, a little thought will make one realize how the opposite happens when the field is reversed. This discovery will make one feel good: giving one the confidence born of success! A victory upon the playing field of science!]

#35 The Reason Why Electrons "Orbit" Protons

> One of the great triumphs of 20th Century science was the discovery by the great experimental scientist Lord Rutherford that the nucleus contains most of the mass of the atom and that electrons "orbit" around it. Today we know that the idea is much more complex. The nucleus is comprised of protons and neutrons and the motions of electrons are confined to "shells". Yet in the midst of all this "vast" knowledge about the atom, nobody in the world knows why electrons are even moving! [At least that was until now!]

The spherical region of less dense space around the proton, and the spherical region of dense space around the electron are responsible for the motion of the electron around the nucleus. It happens like this:

When a proton captures an electron and forms a hydrogen atom, the less dense space surrounding the proton and the denser space surrounding the electron try to bend into each other. This addition changes the density of each region. However, because of its location, the denser space in back of the electron cannot fully blend in. Instead, the sphere of denser space surrounding the back of the electron tries to bend outward from behind the electron and into the sphere of less dense space surrounding the proton or nucleus.

As this dense region tries to bend to one side of the electron, it distorts the electron into a "pear" shape with the pointed end facing the direction the less dense space it is trying to bend towards. And again, just like the pear shapes formed in a "gravitational field", the opposite side of this three dimensional hole called the electron bends towards the side with the sharp bend, moving this side of the hole towards its opposite side, creating motion. However, in this instance, because the electron is so close to the proton, the distortion of the space around the electron is much greater, making the "pear" shape much larger than the pear shape created in the gravitational field. Consequently, the acceleration is enormously large in comparison to that of a gravitational field, and the electron moves at a fantastic speed around the proton.

It is also important to understand that in a hydrogen atom, since the space behind the electron can try to bend outward from around the electron from any angle, the trajectory of the electron around the proton cannot be predetermined. In an atom more complex than the hydrogen atom, the motions of the dense regions of space surrounding other electrons, plus the presence of the additional less dense space surrounding neutrons, all combine to affect the shape of the electrons orbit and the speed of its trajectory.

#36 God! Have We Discovered God!

> *Although this discovery was listed almost last it is not last by any means. It was listed towards the end because it is the great and stunning conclusion of the Vortex Theory: a conclusion that is both shocking and triumphant.*
>
> *When the ramifications of the Vortex Theory were all worked out, shockingly, it was realized that the existence of God is now a real possibility! For everyone who believes in God, this stunning revelation of science is one of its greatest triumphs. The circle is now complete. When science broke away from religion to discover the truths of the universe, one of the truths it discovered was the scientific proof that the existence of God is a real possibility!*

Perhaps the ultimate irony of all ironies occurs when the Vortex Theory is applied to the agnostic viewpoint of the universe. From the agnostic viewpoint of the universe, [to those who do not know if God is real or not], it now has to be concluded that the existence of God is a real possibility! Here's why:

From the agnostic, and atheistic viewpoint of the universe, the consciousness of men is a function of matter. It is merely a function of neurons firing in the brain of the individual. However, matter does not exist in the way it appeared to exist when these beliefs came into being.

The protons, electrons, and neutrons that all the matter in the universe is made of - are really three dimensional holes existing *within* the substance space is made of. Because of this fact, the physical bodies of men are nothing more than a massive collection of holes existing within this same substance. They do not have a separate existence apart from this substance. Nevertheless, these holes have obtained consciousness. *Consequently, if a collection of holes within this substance can obtain consciousness, it is also possible that the substance the holes are made of could gain consciousness of itself too! And if it has, and if it set space into motion, creating the physical universe, then it is God!*

Because of this logic, the existence of God is now a real possibility that must be recognized by science. Anyone who does not acknowledge that the possible existence of God is a real possibility is now acting illogical and irrational!

#37 Universal Religion

> With the discovery that the existence of God is a real possibility, the second discovery is just as awesome; that if there is intelligent life elsewhere in the universe, and if they too discover the true vision of the universe, they will also come to the same conclusion. They will believe in the possible existence of God, and they will know that, if there is other life elsewhere in the universe who also discover the true vision of the universe, that they will come to the same conclusions as them and will be their brothers.

Any form of life in the universe possessing intelligence, and logical reasoning, and clever enough to discover the true vision of the universe, will also come to the conclusion that the substance the universe is made of might also possess consciousness. Like us, they will also realize that if the universe possesses consciousness, and if this consciousness set itself into motion creating the matter of the physical universe, then, it is God.

Consequently, any intelligent creature in the universe that discovers the ultimate scientific vision of the universe will also discover the ultimate philosophical and religious knowledge of the universe. They might not call God – "God", they might use a similar meaning word or title: such as supreme creator; supreme creature; or supreme intelligence: but it will all mean the same thing!

PART III
ADDENDUM
BLOCKBUSTER DISCOVERY!!!

> Note, just prior to publication, an enormous "one of a kind" blockbuster discovery of critical importance was made. For the past 100 years, tens of thousands of scientists including almost every Nobel Prize winning physicist, chemist, and mathematician has tried and failed to solve the meaning of a strange and mysterious number that has come to be called the Constant of Fine Structure – a dimensionless number whose value is 1/137! Here at last, using the principles of the Vortex Theory of Atomic Particles, this seemingly impossible problem was finally solved! It is a twofold achievement because not only is the Constant of Fine Structure explained; it also is another proof of the existence of the two flowing vortices of space between electrons and protons in atoms.

The Explanation of the *"Constant of Fine Structure"*

Since its discovery by Arnold Sommerfeld in 1916, the Constant of Fine Structure has been one of the great unsolved mysteries of science. Perhaps it would not be that important to the world's scientists if it had been discovered by anyone other than Dr. Sommerfeld. Dr. Sommerfeld was no ordinary scientist. He trained many of the great physicists of his day; four of which went on to win the Nobel Prize, including Werner Heisenberg, and Wolfgang Pauli.

Therefore, it seems curious to today's scientific community that neither Sommerfeld nor any of his most famous students could solve the mystery that has come to be called the Constant of Fine Structure – a dimensionless number whose value is 1/137!

It is also shocking to learn that not only Sommerfeld and his group of famous PhD students, but many of the other most famous 20th Century scientists have tried and failed to solve the mystery of the Constant of Fine Structure: Arthur Eddington, Carl Jung, I. J. Good, and Stephen Hawking to name a few. However, perhaps it was Richard Feynman who explained it best when he stated…

"It has been a mystery ever since it was discovered more than fifty years ago, [i.e., actually 100 years now] and all good theoretical physicists put this number up on their wall and worry about it. Immediately you would like to know where this number comes from: is it related to pi or perhaps to the base of natural logarithms? Nobody knows. ***It's one of the greatest damn mysteries of physics: a magic number that comes to us with no understanding by man***".

Nevertheless, the efforts to find a solution to this conundrum have not ceased. Numerous "Quests" to find a mathematical solution to this strange numerological constant continue today. However, the failure of the great 20th Century physicists who have tried and failed to solve this problem for the past 100 years is not their fault. The answer to the mystery of what the Constant of Fine Structure represents has eluded all who have attempted to solve it because they possessed an incorrect vision of how the matter, space, time, energy, and the forces of nature of the universe are constructed.

This discrepancy is cleared up in a scientific paper called, *"The Bases of the Vortex Theory"*. In it is revealed the new and revolutionary discovery that time itself does not exist. Instead, time is discovered to be a function of motion, a phenomenon created by motion; and even more shocking, that everything else in the universe is not made of particles, as presently believed, but instead is constructed out of space: space itself. Unlike the mistaken belief that space is made of nothing, the evidence is now overwhelming that space is indeed made of something: nor is this a return to the old Aether Theory where it was believed that space was made of something and matter was made of solids, [much like the relationship between water and ice which was probably where the idea originated]. And then there is the explanation of the Constant of Fine Structure.

The most important discovery relating to the Constant of Fine Structure is the creation of matter: especially protons and electrons. Already presented in Books 1-3, and in the PhD thesis, was the revelation that particles of matter are really tiny three dimensional [3d] holes in space connected by 3d vortices in 3d space, and fourth dimensional vortices in [4d] space.

These vortices are the key to discovering the explanation of the Constant of fine Structure. But first, we must review 20th Century Science's mathematical discovery of the velocity of the electron "revolving" around a proton in a hydrogen atom in Problem #1 below:

Part 1: The following is according to classical 20th century physics…

Problem #1

Twentieth Century scientists discovered the velocity of the electron in a hydrogen atom using the following mathematics and angular momentum…

Angular momentum = L

Since: $L_1 = I\omega$ where I = moment of inertia = (mr^2); ω = angle of velocity = (v_e/r)

Hence:

$L_1 = (mr^2)(v_e/r) = mrv_e$ where: L = Angular momentum; m = mass of the electron;
   r = radius of the hydrogen atom; and
   v_e = velocity of the electron

However, for the small objects like the electron, L is quantized, so: L also equals:

$L_2 = n\hbar$ Where n = integer [1, 2, 3, etc.]

Because $L_1 = L_2$

Then: $mrv_e = n\hbar$

And: $v_e = n\hbar / mr$

Therefore:

$V_e = n\hbar/mr = \dfrac{(1)(1.055 \times 10^{-34} \text{ Js})}{(9.11 \times 10^{-31} \text{ kg})(0.53 \times 10^{-10} \text{ m})} = 2.19 \times 10^6 \text{ m/s}$

Dividing v_e by "c", the speed of light gives us: $\dfrac{2.19 \times 10^6 \text{ m/s}}{3.0 \times 10^8 \text{ m/s}} = 1/137$

In relation to the speed of light "c": the velocity of the electron is **$v_e = [1/137]$ c = c /137**

Furthermore, according to classical 20th Century Physics, in Figure 1 below, in the hydrogen atom, the proton and the electron are connected by lines of flux from their electrostatic charges; the electron begins to "rotate" about the proton at the incredibly fast speed of c/137. [However, this vision of the atom is a mistake! We will address this problem shortly in Part 3.]

Figure 1

Proton ●═══════════════● Electron ↑ v_e

Part 2: Sommerfeld's *constructed* equation…

As reported in the beginning of this chapter, Arnold Sommerfeld introduced this famous constructed equation of Fine Structure in 1916. It is called "constructed" because it was not derived but was pasted together out of existing constants using the "trial and error method". It consisted of four constants and the number **4π**: the charge of the electron **e**; Planck's modified constant **ℏ**; the speed of light **c**; the permittivity of space **ε₀**; and **4 π**. He put them all together to form an equation:

Equation #1: $4 \pi \varepsilon_0 \hbar c \alpha = e^2$

It is also interesting to note that he put the subject of his equation Alpha [α] in the body of his equation rather than as the subject of his equation. Consequently, it has since been modified:

Equation #2: $\alpha = e^2 / 4 \pi \varepsilon_0 \hbar c$

And, when **α** is solved for in the above mathematical formula, all of their units cancel out, leaving only a dimensionless number where: $\alpha \approx 1/137$.

A. Finding the value of the Constant of Fine Structure using traditional 20th Century methods…

To examine Sommerfeld's equation we must first define the numbers and units he used. These numerical values and the physical units of the four constants he used in the Constant of Fine Structure are as follows:

#1. The Electron's charge: $e = 1.60217653 \times 10^{-19}$ C; where C = Coulombs

#2. Planck's modified constant: $\hbar = [6.626 \times 10^{-34} \text{ kg m}^2/\text{s}] / 2\pi$

#3. Speed of light: $c = 3.0 \times 10^8$ m/s

#4. Permittivity of space: $\varepsilon_0 = 8.8541878176 \times 10^{-12}$ s² C² / m³ kg

---

B. Eliminating the units of the four constants used in Equation #2...

In Sommerfeld's famous equation, it was discovered that the units of all four constants canceled each other out leaving a dimensionless number of 1/137; as seen in Problem #2 below:

Problem #2
Modifying equation #1 to equate it to alpha α instead of e^2 gives us:
Equation #2 $\alpha = e^2 / (4\pi)(\varepsilon_o)(\hbar)(c)$

Contemporary science's view of Equation #2 adding units and numbers gives us Equation #3:

$$\alpha = \frac{(1.60 \times 10^{-19} \text{ C})^2}{(4\pi)(8.854 \times 10^{-12} \text{ s}^2\text{C}^2/\text{m}^3\text{kg})([6.626 \times 10^{-34} \text{ kg m}^2/\text{s}]/2\pi)(3.0 \times 10^8 \text{ m/s})}$$

Using 20th Century logic, all of the units **C, m, s, & kg** then cancel out as follows:

$$\alpha = \frac{(1.60 \times 10^{-19} \text{ C})^2}{(4\pi)(8.854 \times 10^{-12} \text{ s}^2\text{C}^2/\text{m}^3\text{kg})([6.626 \times 10^{-34} \text{ kg m}^2/\text{s}]/2\pi)(3.0 \times 10^8 \text{ m/s})}$$

$(4\pi / 2\pi) = 2$; C^2 cancels C^2; m^3 cancels m^2 & m; s^2 cancels s & s; and **kg** cancels **kg**. Leaving us with no units and only the following numbers:

$$\alpha = \frac{(1.60 \times 10^{-19})^2}{(2)(8.854 \times 10^{-12})(6.63 \times 10^{-34})(3.0 \times 10^8)}$$

When these numbers are multiplied and divided together, we get:
$\alpha = 7.276 \times 10^{-03} = 1/137.4 \approx 1/137$

Part 3: 20th Century physics vs. *The Vortex Theory of Atomic Particles*...

Returning our attention to Figure 1, we see that the Bohr model of the hydrogen atom views the electron as "rotating" around the proton at a velocity of c/137 [v_e].

Figure 2

Proton ●━━━━━━━━━━━━━━━━━━● Electron   ↑ v_e

The above vision of the hydrogen atom is also seen in this simplified schematic drawing below:

Figure 3

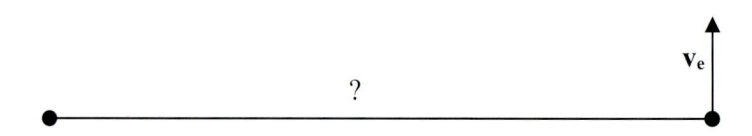

●─────────?─────────●   ↑ v_e

The question mark was used because 20th Century physicists did not know what was happening between the electron and the proton! They assumed that the electrostatic charges between the electron and the proton are static and not moving. In fact, this is why these "charges" are called "electro*static*" charges! The very belief that these charges are unmoving lines of flux is what elicited this name *static* in the compound word - "electro-*static*"! But this is not so!

The *Vortex Theory of Atomic Particles* [in Books 1-3] reveals that the electrostatic charges of the proton and the electron are created by 3d space flowing into and out of them.

A. Background information: the construction of protons and electrons…

The *Vortex Theory of Atomic Particles* discovered that protons and anti-protons are three dimensional (3d) holes in space connected in fourth dimensional space by fourth dimensional (4d) vortices. In Figure 4 below, P = the proton; while $\overline{P}$ = anti-proton:

Figure 4

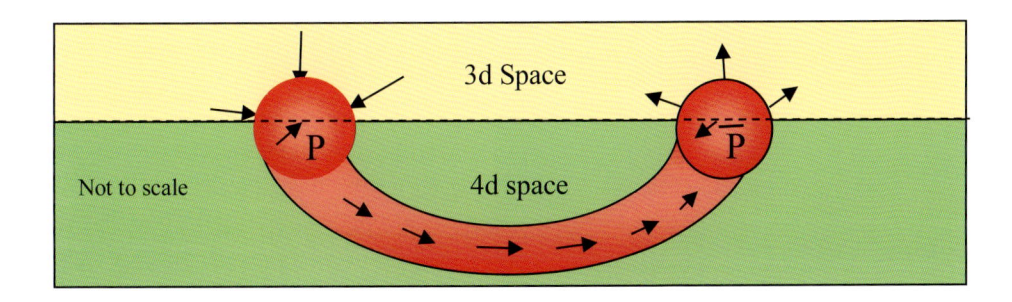

As 3d space flows into the proton, the surrounding 3d space becomes less dense. As space flows out of the anti-proton, the surrounding 3d space is pushed outward making it denser. The same is true for the electron (e) and its anti-particle, the positron ($\overline{p}$).

Figure 5

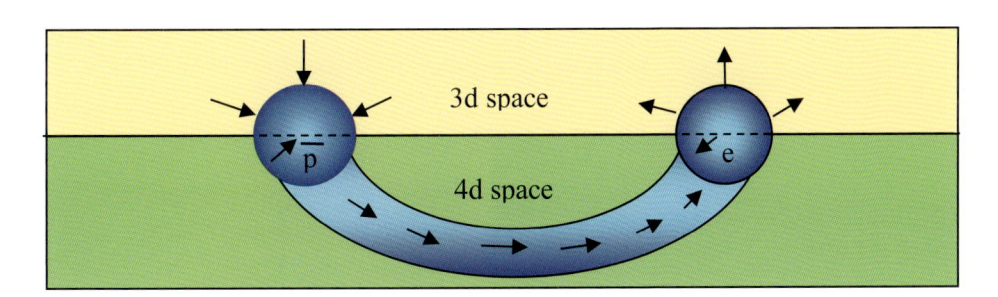

Like the proton and anti-proton, space flows into the positron ($\overline{p}$) making the volume of space surrounding it less dense; and, as space flows out of the electron (e), it pushes the surrounding space outward making it denser.

B. Creation of the hydrogen atom…

When the electrostatic forces of the proton and the electron draw these two "particles" towards each other; the electron begins to "rotate" about the proton and their vortices in 4d space cross. When the vortex connecting the proton and the anti-proton, and the vortex connecting the positron and the electron cross in 4d space, their vortices break and reconnect. The result causes the proton to now be connected to the electron, and the positron to be connected to the anti-proton, creating a hydrogen atom [and somewhere in the universe, an anti-hydrogen atom].

Figure 6

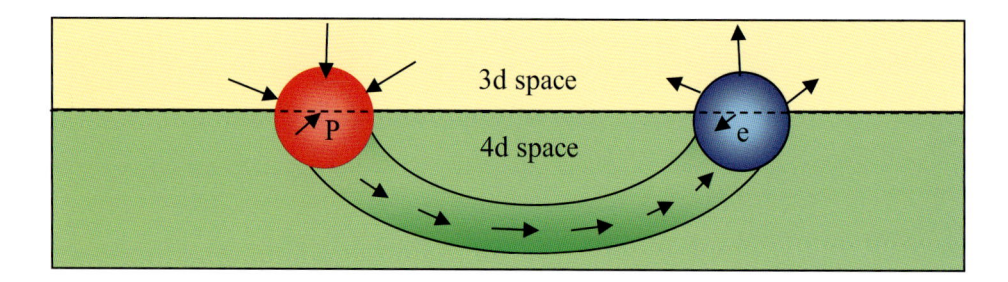

As these two holes move closer together, a critical distance is reached where all of the 3d space flowing out of the electron flows directly into the proton. A second vortex now flows from the electron back to the proton in 3d space as seen in Figure 7 below:

These two vortices create a circulating flow containing a fixed volume of space. This circulating volume of 3d space continually flows from the proton into 4d space, through 4d space, then back into the electron. Here, it exits the electron, flowing back through 3d space into the proton once again, binding the proton to the electron creating a hydrogen atom.

Figure 7

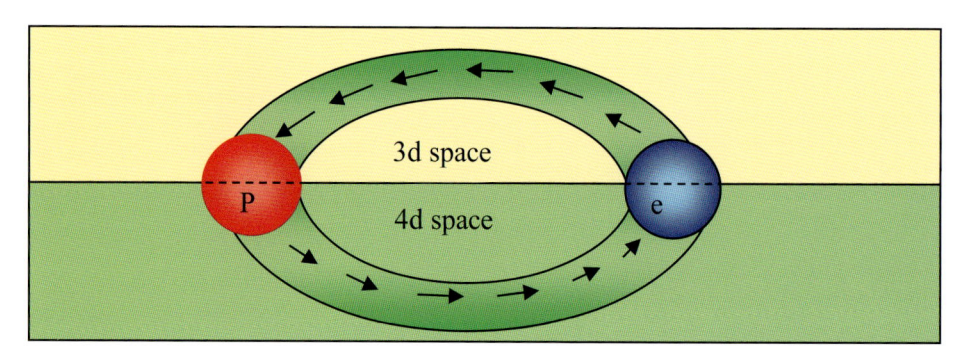

When the circulating flow commences, both of the electrostatic charges are neutralized. The word "neutralized" was used because no flowing space escapes from the system. If surrounding space still flowed into or out of this system, atoms would possess electrical charges.

Because the electrostatic charge of the proton and the electron are equal, the two volumes of 3d space flowing into the proton and out of the electron are equal. This means that the volume of the 3d space flowing out of the "three dimensional surface area" of the electron is exactly the same as the volume of the 3d space flowing into the "three dimensional surface area" of the proton.

Since these surface areas are of greatly different sizes, [but still possess the same volume of 3d space flowing through each], the density of the vortex is different at each end.

Coming out of the electron, the volume of 3d space is condensed [it is *denser*], while the volume going into the proton is the same but *less dense*. This phenomenon can be expressed using imaginary lines. For example, blowing up each end to enormous sizes, reveals that the imaginary lines going into the proton are much further apart than the lines coming out of the electron; showing the space is *denser* coming out of the electron and *less dense* going into the proton.

Figure 8   The vortex in (3d) space flowing from the electron to the proton in a hydrogen atom.

Proton   Electron

Part 4: When the rotation begins…

 The schematic drawing of the 4d vortex flowing from the proton to the electron in 4d space is shown below.

Figure 9

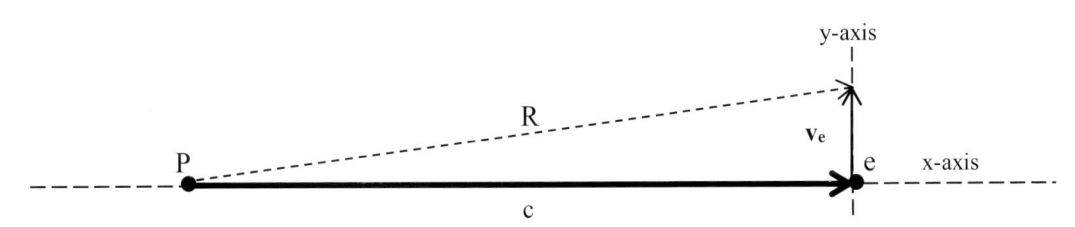

y-axis

R

Ve

P   c   e   x-axis

   Just looking at the above drawing suddenly creates a problem. According to the above schematic, using vector addition, the velocity of the space in the vortex is no longer traveling at the speed of light c, but instead is now traveling at the higher velocity of the resultant vector R. This cannot happen; if it did, photons ejected from the hydrogen atom would exceed the speed of light. We have never seen such a phenomenon; all photons travel at the speed of light c. Therefore, the velocity of the 3d flowing space in the vortices has to drop and start moving at the slower Apparent velocity of "Av".

Figure 10

[The 4d velocity of the vortex flowing from the proton to the electron is shown below in red.]

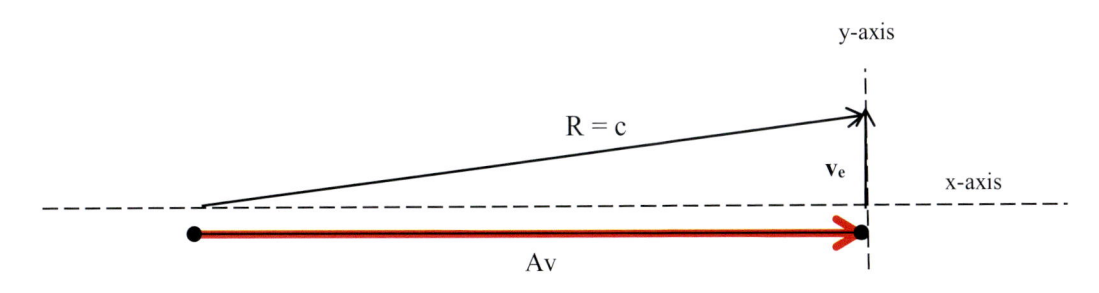

y-axis

R = c

Ve   x-axis

Av

(Where Av = Apparent velocity)

   Consequently, *when the electron is "rotating"*[2] the velocity of the space in the vortex is no longer traveling at the speed of light c; instead, it is now traveling at the slower Apparent velocity of Av. To determine what the Apparent velocity is we turn our attention to Problem #3 below:

---

[2] Note, when the atom absorbs or expels a photon, the electron <u>stops rotating</u>, and the velocity of the space in the vortices jumps back up, reaching the velocity of the speed of light again: c.

Figure 11

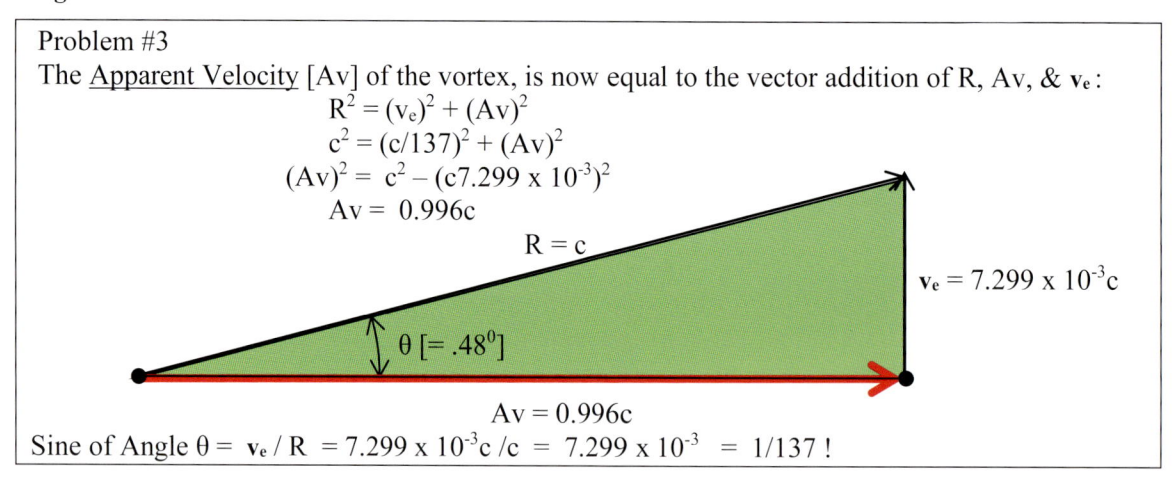

> Problem #3
> The Apparent Velocity [Av] of the vortex, is now equal to the vector addition of R, Av, & v_e:
> $$R^2 = (v_e)^2 + (Av)^2$$
> $$c^2 = (c/137)^2 + (Av)^2$$
> $$(Av)^2 = c^2 - (c7.299 \times 10^{-3})^2$$
> $$Av = 0.996c$$
>
> R = c
> $v_e = 7.299 \times 10^{-3} c$
> θ [= .48°]
> Av = 0.996c
>
> Sine of Angle θ = v_e / R = $7.299 \times 10^{-3} c$ /c = 7.299×10^{-3} = 1/137 !

Consequently, the Constant of Fine Structure α, can be defined as the difference between the actual velocity of the flowing space in the vortices, and their apparent velocity! This difference is equal to the sine of angle θ between them: 1/137! The Constant of fine Structure is dimensionless, because the sine of an angle is dimensionless!

After a hundred years of analysis, frustration, and speculation – the above seemingly modest explanation for the Constant of Fine Structure suddenly seems wholly inadequate. But this is not so. The above explanation has absolutely shocking, life changing, and universal implications not only for all of physics, chemistry, and astronomy; but also, for all carbon based life forms in the universe: including us! Without the Constant of Fine Structure, we would not be here! Here is why…

Part 5: *All life in the universe is dependent upon the Constant of Fine Structure!!!*

Without the element carbon, there would be no life in the universe; but without the Constant of Fine Structure, there would be no carbon! The Constant of Fine Structure is essential for the creation of the carbon atom found in all living creatures: including us! Amongst other things, the Constant of Fine Structure creates what has come to be called a "coupling constant" between protons in carbon atoms; allowing the nuclei of less massive atoms to bond together, allowing them to create the larger carbon nucleus. Its importance cannot be overstated…

In 1954, Sir Fred Hoyle, a British astrophysicist, realized that if electromagnetism were only a little stronger, stars could not create carbon. More recent studies indicate that if the fine-structure constant were 2 to 2.5% higher, the universe would be carbon-free, and there would be no chemical basis for life! So how does the Constant of Fine Structure affect the production of carbon in stars?

One of the great unheralded discoveries of the Vortex Theory of Atomic Particles is that the slower Apparent velocity of the 3d space flowing in the vortices, also slows the volume of 3d space flowing into and out of the proton. This slows its rotation, reducing the strength of its magnetic moment; reducing its ability to repel the magnetic fields of other protons in other light nuclei [whose magnetic moments have also decreased]; making it easier for the intense pressures within stars to compress these lighter nuclei together [to couple], to bond to form more massive nuclei; making it easier to form carbon in stars.

> The slower apparent velocity of the space flowing in the vortices, is not only responsible for creating the Constant of Fine Structure, it also changes the magnetic moment of the proton. This is not speculation but is based upon the spin magnetic moment of a Dirac particle μ_N; of charge e; with the reduced Planck constant ℏ; and with a proton's mass of m_p:
>
> Equation #3 $\qquad \mu_N = e\hbar / 2m_p$
>
> Where e = elementary charge; ℏ = reduced Planck constant; m_p = proton's mass.
>
> Because the magnetic moment of the proton μ_p is equal to 2.79 μ_N, then: $\mu_p / 2.79 = \mu_N$
>
> Equation #4 $\mu_p / 2.79 = e\hbar / 2m_p$ or; $\mu_p = 1.395\, e\hbar / m_p$
>
> Consequently, we see that the magnetic moment of the proton μ_p is directly proportional to e, the charge of the proton.

Because the magnetic moment of the proton μ_p is equal to 2.79 μ_N; we see that the proton's magnetic moment is directly proportional to μ_N in Equation #3. Consequently, the magnetic moment of the proton in the nucleus of an atom is then directly proportional to [e^+] the charge of the proton in Equation #3. Therefore, when the velocity of the electron about the proton causes the space flowing in the vortices to now move at the slower Apparent velocity [Av]; this decreases the volume of the space flowing into and out of the proton. This decreases its charge [e^+]; causing the proton's magnetic moment [μ_p] in Equation #4 to decrease. This then allows the protons in the nuclei of the atom to couple more easily with the nuclei of other atoms, whose magnetic moments have also decreased, and are also trapped within the intense pressures inside the interiors of stars; allowing three of them to "couple", to bond; to create a more massive nuclei, allowing the creation of the carbon atom.

It is suddenly apparent that this seemingly insignificant little number of 1/137 has a profound effect upon all of our lives: and life itself! It reveals a simply beautiful yet delicate relationship between the creation of the physical matter in the universe, and the creation of all the biological life in the universe! WOW...what a wonderful universe we live in! God did a good job!

DISCUSSION

The triumph of discovering the explanation for the mysterious Constant of Fine Structure dimensionless number of 1/137 comes not without consequences. The answer reveals that science's vision of the electrostatic charges surrounding subatomic particles is wrong. They are not "static" charges at all. Instead, they are dynamic, and should be called "electro-dynamic" charges because they are constructed out of three dimensional space flowing into and out of the "particles" of nature; nor are the particles of nature particles. Instead, they are tiny three dimensional holes created upon the three dimensional surface of fourth dimensional space.

Without this vision of the tiny three dimensional holes, and the equally tiny two vortices of flowing space, the Constant of Fine Structure cannot be explained.

This same vision is crucial for the creation of the carbon atom in stars. Again, the presence of the two miniature vortices of flowing space between protons and electrons is vital for this explanation. The speed of the electron moving about the proton forces the velocity of the flowing space in the vortices to drop below the speed of light and flow at a slower Apparent velocity. This drops the volume of space flowing into and out of the proton, lowering its electric charge; lowering its magnetic moment, allowing it to couple with other lighter atoms within the interior of a star that are also experiencing the same lowered magnetic moment; allowing the formation of the carbon nucleus. None of which could occur without the presence of the two vortices flowing into and out of these tiny three dimensional holes.

Consequently, the discovery of the explanation for the Constant of Fine Structure is as much of a discovery of it, as it is another proof of the Vortex Theory of Atomic Particles. Science cannot have one without the other.

> Note: the following discoveries all flow from the discovery of the explanation for the Constant of Fine Structure. They have been put into what is being called:
>
> Ch 2A Creation of the 21 centimeter line
> Ch 3A The explanation of the twin fine lines of hydrogen
> Ch 4A "Electrostatics" is a mistake: it should be "Electrodynamics" !!!
> Ch 5A "Electrodynamics" explains the ejection of alpha particles, gamma rays, x-rays, and electrons

Chapter 2A Creation of the 21 centimeter line

One of the most important discoveries ever made in astrophysics was the discovery of the 21 centimeter line. This seemingly insignificant little line on a light spectrum photograph allows us to identify the presence of hydrogen in our Milky Way Galaxy and the universe.

The tiny infrared photon's wavelength that creates the 21 centimeter line was first observed in 1951 in light spectra by Harold Ewen and Edward Purcell of Harvard University. The photon's wavelength is now used to identify the presence of hydrogen in dust clouds allowing astronomers to see the arms of the Milky Way Galaxy, and form theories about star creation using hydrogen.

The explanation of how this tiny photon is created allows us to explain a whole host of other phenomenon in the universe; some of which are unknown to today's science; the one of most importance – that is of vital importance – is the subject of this chapter. Its explanation begins with a strange discovery made in 1959 by a brilliant mathematician named Dr. Stephen Smale.

In 1959, Dr. Stephen Smale, a PhD in mathematics and a mathematical genius, published his famous paper entitled: *A Classification of Emersions of the Two Sphere*. Unless one is also a genius in mathematics, neither the title, nor the paper – literally stuffed with abstract mathematical formulas and theorems – seems to make any sense at all. However, to all other mathematical geniuses it presented a simply astounding discovery: that a three dimensional sphere can be turned inside out without tearing its surface!

It was only a few years later, when the mathematics of this paper was used to create visual images, that people could see and appreciate what Dr. Smale had done. These images allowed the drawing of pictures that created a sensation when they were finally published in Scientific American Magazine in 1966.

Years ago, when it was first encountered, Dr. Smale's mathematical analysis of a sphere's ability to turn inside out seemed like nothing more than a strange curiosity developed by a brilliant mathematician with a whimsical imagination: something that has no practical purpose whatsoever. However, all that has suddenly changed with the discovery of the *Vortex Theory of Atomic Particles*.

The work of Dr. Smale is suddenly, and absolutely critical in explaining many of the strange phenomenon of the universe: such as the creation of neutral [no electrostatic charges] subatomic particles like the neutron; the creation of the Fine Lines of Hydrogen; and plus, nothing less than the creation of the universe itself! This will be revealed at the end of this book, but first we need it to explain how an electron circling a proton in a hydrogen atom, that is subsequently struck by another electron in another hydrogen atom – ends up turning inside-out! The following little experiment helps…

Part 1: When left becomes right, and right becomes left!

To better understand what happens when a three dimensional hole turns inside out, try this little experiment. Get an old T-shirt and a permanent black felt ink pen. Then on the right sleeve draw an arrow that points towards the front of the shirt and mark an "R" beside it; [do it slow enough so that the ink bleeds through to the other side of the shirt], on the left sleeve draw an arrow that points towards the back of the shirt and put an "L" beside it. Now put the shirt on [if you draw the arrows and letters with the shirt on, the ink will bleed through and get on your skin.] When done it will look something like the figure below:

Figure 1 [looking from the top down:]

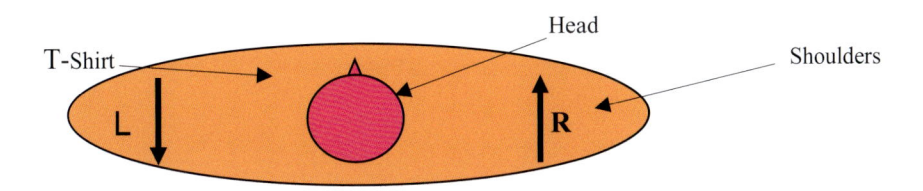

Looking at the above figure, it is easy to see that if one follows the arrow's directions and slowly turns around rotating, one will end up rotating COUNTERCLOCKWISE.

However, when the T-shirt is taken off, turned inside out, and put back on, a different situation arises:

Figure 2 [looking from the top down:]

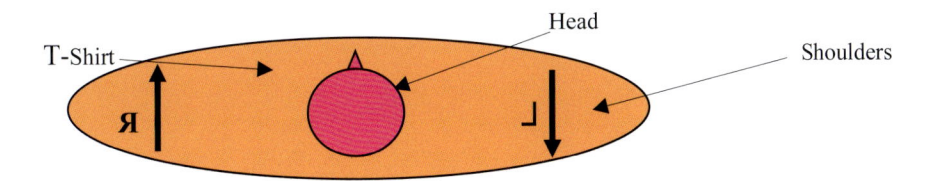

Suddenly everything is reversed. Not only what was on the left is now on the right, and on the right is now on the left, but the arrow on the left now points forward and the arrow on the right now points backwards. If one again follows the directions of the arrows and rotates, the rotation is now CLOCKWISE not counterclockwise! The exact same situation occurs when the electron is turned inside out as per. Smale's equations[***].

So, what happens when the electron turns inside out in a hydrogen atom? To answer this question, we must return our attention to the construction of the proton, the electron, and how they create the hydrogen atom.

Part 2: The creation of dense and less dense regions of space

In 3d space, the 3d space flowing into the proton creates a volume of less dense space as it is "pulled into the 3d hole that is the proton". Immediately surrounding the electron, the 3d space flowing out of this 3d hole pushes outward creating a volume of denser space [note: Figures 4 & 5 below are not to scale].

In the two drawings below, note how the proton's less dense volume of space surrounding it is much greater than the electron's smaller denser volume of space. It must also be said that the neutron possesses a volume of less dense space surrounding it too. This was originally discovered and presented in the first book in this series called, *The Vortex Theory of Atomic Particles.*

These less dense regions of space are responsible for creating what can only be called Nuclear Gravity[***]. On a larger scale, the additions of billions of trillions of these less dense regions of space create the force of gravity. It is not "Bent Space" that creates gravity, but less dense space; and denser space creates "Anti-gravity! Already proven and published that the denser region surrounding the electron creates an Anti-gravity Force. A force that properly used, will someday take us to the planets and the stars. But we are getting ahead of ourselves. We must understand the creation of the hydrogen atom first.

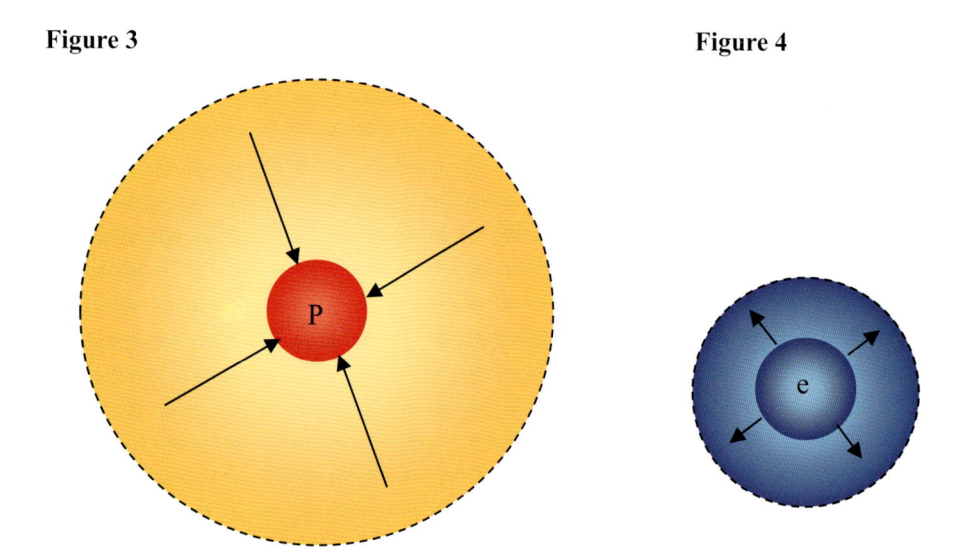

Figure 3

Figure 4

Part 3: Creation of the hydrogen atom

According to the principles of the Vortex Theory of Atomic Particles, when the electrostatic forces of the proton and the electron draw these two "particles" towards each other, they move close; begin to rotate, and in doing so cause their vortices in 4d space to cross. When the vortex connecting the proton and the anti-proton, and the vortex connecting the positron and the electron cross in (4d) space, they break and reconnect. The result causes the proton to now be connected to the electron, and the positron to be connected to the anti-proton.

When a hydrogen atom is created, although present day science says that the electrostatic charges of protons and electrons pull these two "particles" together, this is an oversimplification. What really happens is that some of the space flowing out of the electron begins to flow into the proton; and as these two holes move closer together, a critical distance is reached where all of the 3d space flowing out of the electron flows directly into the proton. When this situation occurs, *a second* vortex of whirling space is created. The second vortex now flows from the electron back to the proton in 3d space.

These two vortices create a circulating flow containing a fixed volume of space. This circulating volume of 3d space continually flows from the proton into 4d space, through 4d space, and then back into the electron. Here, it exits the electron, flowing back through 3d space and into the proton once again, binding the proton to the electron creating a hydrogen atom.

Figure 5

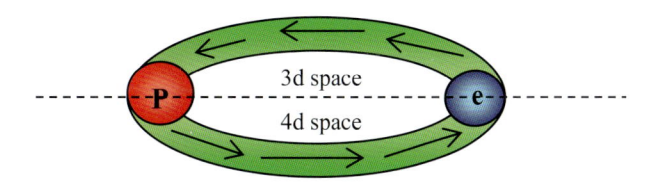

When the circulating flow commences, both of the electrostatic charges are neutralized. The word "neutralized" was used because no flowing space escapes from the system. If surrounding space still flowed into or out of this system, atoms would possess electrical charges.

Because the electrostatic charge of the proton and the electron are equal, the two volumes of 3d space flowing into the proton and out of the electron in 4d space are equal. This means that the volume of the 3d space flowing out of the "three dimensional surface area" of the electron is exactly the same as the volume of the 3d space flowing into the "three dimensional surface area" of the proton. Since these surface areas are of greatly different sizes, [but still possess the same volume of 3d space flowing through each], the density of the vortex is different at each end.

Coming out of the electron, the volume of 3d space is condensed [it is *denser*], while the volume going into the proton is the same but *less dense*. This phenomenon can be expressed using imaginary lines. For example, blowing up each end to enormous sizes, reveals that the imaginary lines going into the proton are much further apart than the lines coming out of the electron; showing the space is *denser* coming out of the electron and *less dense* going into the proton.

Figure 6 The vortex in (3d) space flowing from the electron to the proton in a hydrogen atom.

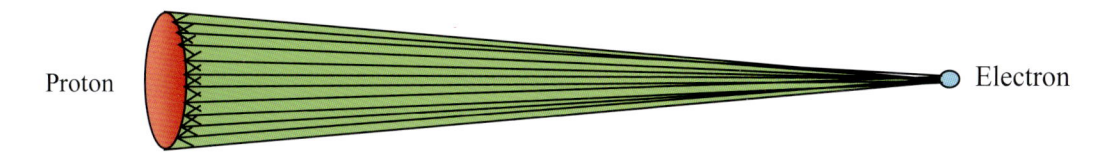

Part 4: Creation of the density gradient & the corridor

In the Vortex Theory of Atomic Particles, it is revealed that 3d space is flowing into the proton making the volume surrounding it less dense; while 3d space flowing out of the electron, pushes outward, making the surrounding space denser. Consequently, the only way space can flow between the two ends, is for them to get close enough to each other so that some of the dense space from the electron and the less dense space surrounding the proton meet, blend into each other, and create a "gradient of denser space" between them. At the electron, the gradient is denser; but as the gradient lengthens out, its density decreases near the proton. However, it is still denser than the normal gradient of space as seen in Figure 8…

Figure 7

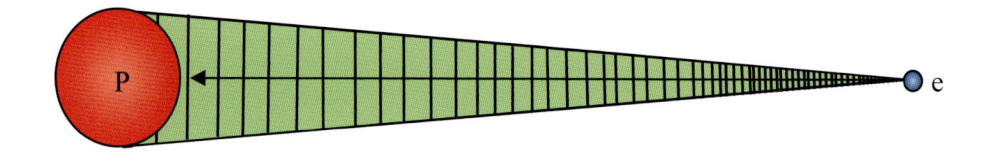

Figure 8

Normal gradient of space:

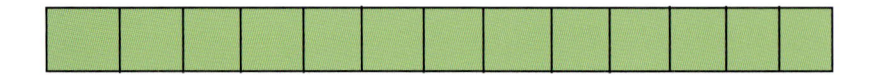

As the proton and electron are drawn together by the Coulomb forces, the formation of the "gradient of space" directly between them creates this "Corridor" of dense space flowing between them as seen in Figure 9. Because this "corridor" of space is denser than the normal less dense volume of space surrounding the proton, it creates the "cone shape."

Figure 9

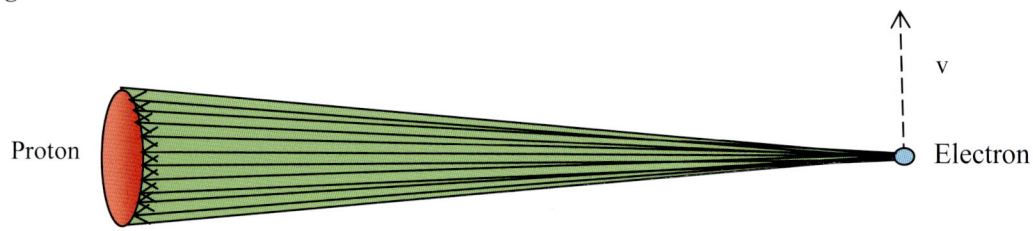

Part 5: The hydrogen atom is hit by another hydrogen atom.

The rotation orientation of protons and electrons in atoms are opposite to each other. This rotation is not only responsible for the explanation of the Pauli Exclusion Principle that this theory discovered, but is also crucial in the explanation of the 21 centimeter line.

This rotation can best be explained using the analogy of a rotating pipe. Say for example that we have a section of pipe that is rotating on a lathe. If we look at one end, the pipe is seen to be rotating in a counter clockwise direction; but when we walk around to the other end, the rotation is clockwise:

Figure 10

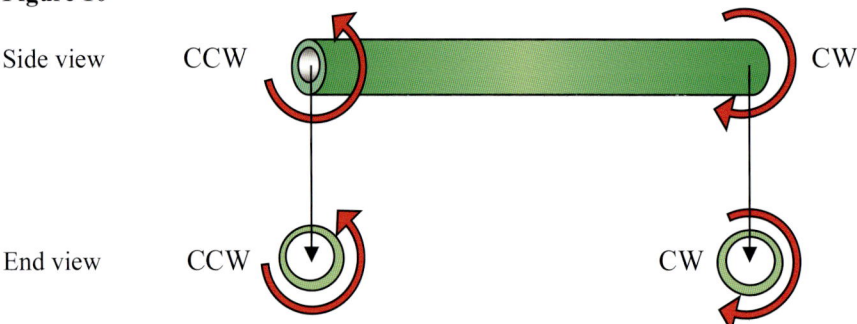

Side view   CCW   CW

End view   CCW   CW

The same is true for the vortices. Because the proton is merely a 3d hole at one end of the 4d vortex [quarks are explained in Book 3], and the electron is a hole at the other end, when one rotates counterclockwise (called Up), the other is rotating clockwise (called Down).

Part 6: The collision

When atoms collide, it is the electrons that actually collide. So, when two hydrogen atoms are perpendicular to each other, and the rotations of the electron are opposite to each other, when the electrons collide, in relation to each other's orientation, one will be spinning CCW; the other CW.

Although it is not known at the time of this writing if the opposite spin of each electron is transferred to the other [only future experimentation will tell], what is known is that during the collision; one transfers its spin to the other, causing it to turn inside out, turning it from a CCW to CW spin or vice versa: causing the electron to change its spin orientation from an up to a down or a down to an up.

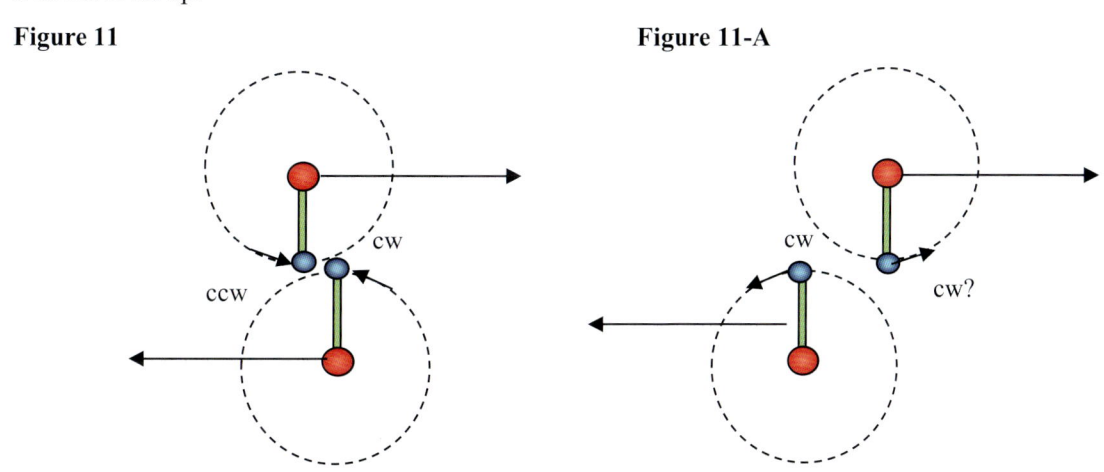

Figure 11   **Figure 11-A**

Because of the concept of a 3d hole turning inside out is almost impossible to visualize, the following drawings using a vortex with an outside color of green and an inside color of red will help. Notice first in the below drawing how the outside color is green, and the inside red:

Figure 12

Left end   Right end

Figure 13 When the pipe rotates

When the pipe is put on a lathe and one stands at the left end seeing it rotate CW; from the perspective of the other end, it will be seen to rotate CCW. This reveals that the spin states are reversed from end to end: if the left end is CW; then the right end is CCW:

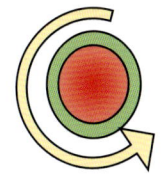

Part 7: Twisted, flowing space.

Now say that the electron was turned inside out. This causes the space flowing in the vortices to be twisted inside out also. To be able to flow back to the proton in 3d space, and from the proton back to the electron in 4d space, both vortices have to straighten back out; causing them to twist in a shape similar to that of a Mobius strip. This twist is illustrated by the purple wave in the tube. Here, on one side of the twist, the inside and the outside of the vortex are turned inside out: the outside (green) is now the inside, and the inside (red) is now on the outside! The 4d vortex is shown below:

Figure 14 Fourth dimensional vortex

Note: the purple squiggle in the vortex represents the twist in its structure similar to that of a Mobius strip:

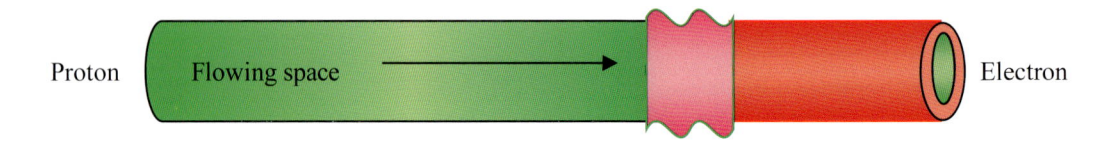

Figure 15

Notice how the spin states of the proton and the electron are now the same: the left, CW; the right CW: this occurs because the electron is also turned inside out:

Figure 16 Three dimensional vortex

Notice how the same situation has also occurred in the 3d vortex in 3d space. It also has a twist in it similar to a Mobius strip.

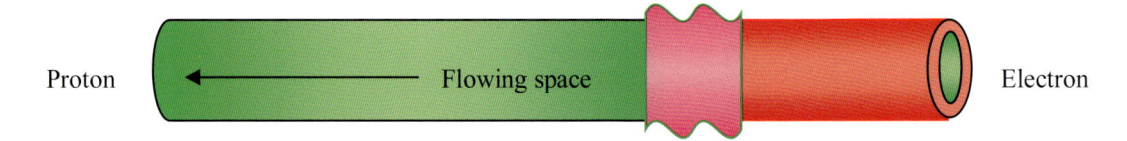

Figure 17

Notice how the spin states of the proton and electron are still the same: the proton CW; the electron CW:

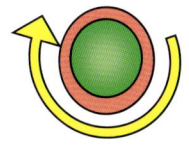

The electron turning inside out causes the vortices to elongate and the electron to reverse its spin. It goes from say spin up to spin down, or from spin down to spin up. Because we saw earlier in Chapter 4 Problem # 1 that for small particles such as electrons, ***the angular momentum is quantized, causing the same amount of elongation for all vortices in all hydrogen atoms in the universe!*** This elongation of the vortices, caused by the unnatural twist in the vortices, equates to an extra volume of flowing space being added to both vortices.

Part 8: The twisting back of the vortices causes the creation of the 21 centimeter line.

Because both of these bends in space are unnatural, eventually they will want to twist back; when this happens, the extra volume of space in the vortex is released as a photon.

This sequence can be seen in the following drawings using the 4d vortex as an example:

Figure 18 Side view
Normal state of the flow in the 4d vortex seen below; the 3d space in the 4d vortex flows directly from the proton to the electron in straight lines. Notice how their spins are reversed:

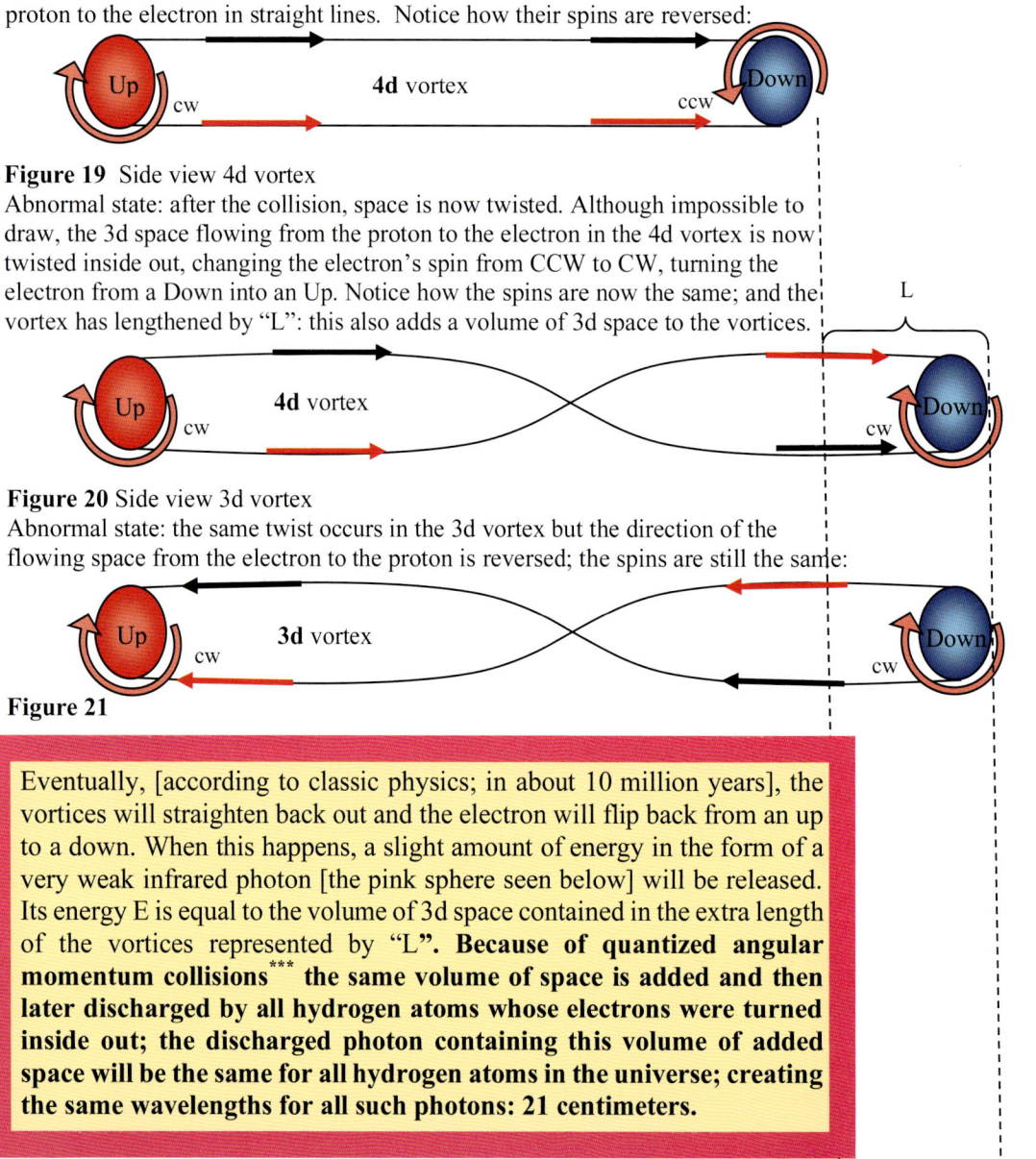

Figure 19 Side view 4d vortex
Abnormal state: after the collision, space is now twisted. Although impossible to draw, the 3d space flowing from the proton to the electron in the 4d vortex is now twisted inside out, changing the electron's spin from CCW to CW, turning the electron from a Down into an Up. Notice how the spins are now the same; and the vortex has lengthened by "L": this also adds a volume of 3d space to the vortices.

Figure 20 Side view 3d vortex
Abnormal state: the same twist occurs in the 3d vortex but the direction of the flowing space from the electron to the proton is reversed; the spins are still the same:

Figure 21

Eventually, [according to classic physics; in about 10 million years], the vortices will straighten back out and the electron will flip back from an up to a down. When this happens, a slight amount of energy in the form of a very weak infrared photon [the pink sphere seen below] will be released. Its energy E is equal to the volume of 3d space contained in the extra length of the vortices represented by "L". **Because of quantized angular momentum collisions*** the same volume of space is added and then later discharged by all hydrogen atoms whose electrons were turned inside out; the discharged photon containing this volume of added space will be the same for all hydrogen atoms in the universe; creating the same wavelengths for all such photons: 21 centimeters.**

122

Figure 22 The 21 cm emission line seen in the light spectrum from a hydrogen dust cloud in the Milky Way Galaxy.

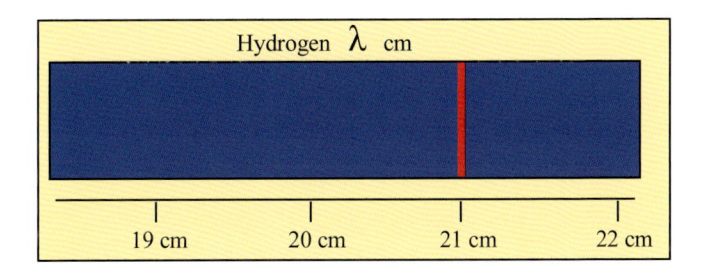

Chapter 3A The explanation of the twin fine lines of hydrogen

Again as mentioned before, during a collision between two hydrogen atoms, because the angular momentum of the electrons are quantized, unlike collisions with other matter in the universe, the amount of angular momentum that is transferred from one hydrogen atom to the other is always the same: creating the same elongation of the vortices in one of the atoms; causing the same radius increase every time. This is a very important concept.

In a massive hydrogen cloud, light years in diameter, because many hydrogen atoms will have collided with other hydrogen atoms, changing electron spins and lengthening the vortices, two sets of photons will now be released from the cloud. For the hydrogen atoms that have vortices of "normal length" (have opposite spin states) and have protons and electrons of opposite spins [seen in Figure 18], these atoms will absorb photons of a certain energy [of a certain volume]. These photons will add to the flowing space in the vortices, causing the vortices to lengthen, and allowing the electron to jump up to its next energy state, staying there until the electron finally falls back to its ground state; and the extra volume is discharged back into the universe in the form of a photon.

However, for "abnormal" hydrogen atoms that have electrons that are turned inside out, when both the proton and the electron have the same spin states because of the abnormal twist in the vortices [seen in Figure 19], this causes the vortices to lengthen slightly. When this happens, and becomes for all practical purposes the new ground state of the hydrogen atom, it now takes a slightly more powerful photon [larger volume] to make the electron jump to its next energy state and stay there; and when the vortices again finally shorten, returning the electron again to its new ground state, this slightly more energetic photon is released into the universe.

Figure 23

Expansions and contractions of a photon with more volume is seen below. Note: the frequency of increases, and the wavelength λ decreases when compared with Figure 7. [Note: pink is exaggerated extra volume of the photon:

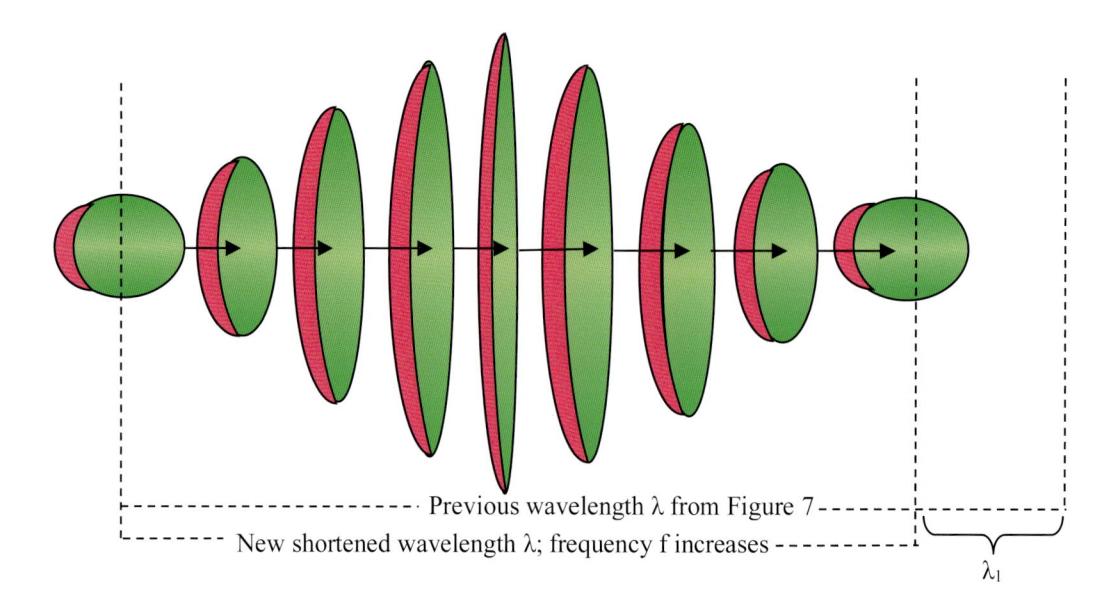

·········· Previous wavelength λ from Figure 7 ········
‒‒‒‒‒ New shortened wavelength λ; frequency f increases ‒‒‒‒‒
λ_1

Where $f_1 = c \lambda_1$ the separation frequency of the fine structure lines in the spectrograph of hydrogen seen in Figure 24 below.

[The photon can now be defined as a packet of condensed three dimensional space, expanding and contracting perpendicular to its direction of travel.]

When a camera is aimed at the cloud of hydrogen, and the photograph taken, two different energy states of hydrogen are now seen: the first is from the "normal" hydrogen and the second from the "abnormal" hydrogen. The difference in the volumes of these two photons of light are small, making the difference in their vibratory frequencies so small [f_1 in Figure 16], that only when the light is spread out by a spectrograph do they create two separate lines: the Fine Structure lines first seen by Albert Michelson and his colleague Edward Morley.

Figure 24

Double lines of hydrogen create the constant of fine structure.

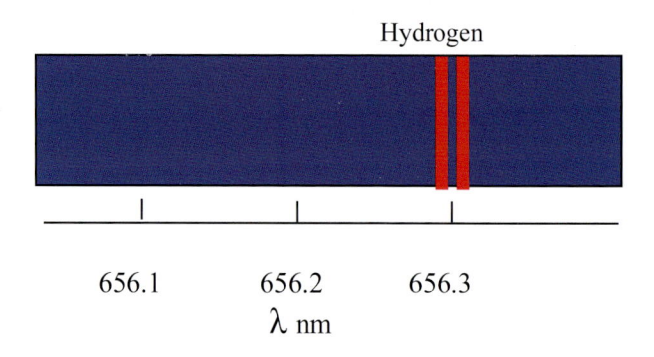

Hydrogen

656.1   656.2   656.3
λ nm

Chapter 4A "Electrostatics" is a mistake: it should be "Electrodynamics"! ! !

One of the unpleasant duties of discovering revolutionary knowledge is the discarding of old ideas: ideas that in many instances we have come to accept as true and have grown comfortable with. The more comfortable we are, the harder it is to let go when this knowledge is suddenly and unexpectedly revealed to be mistaken. Such was the situation in days of old when men were faced with the knowledge that the earth was not flat but round. A traumatic experience for 15th Century sailors who were afraid of falling off the edge of the earth. When Columbus boldly tested this idea by sailing west to the east, the sailors on his ship were so afraid of falling off the edge of the earth, they almost mutinied and hung Columbus. Luckily for us he proved his revolutionary idea was right by sighting land and they backed off.

Although [hopefully] we are no longer forced to confront life and dearth struggles by proposing new ideas, the stress they create when first introduced is still real. However, it has to be done if we are to progress onward into the future. For example…

The word Electrostatics: is a mistake.

There is an unseen war going on in the minds of each and every human being on this planet. Although not talked about or even perceived; it is a war between our perception of reality and reality itself. When our ancient ancestors looked up at the sky and observed the motions of the sun, moon, stars and planets, it seemed perfectly logical to assume everything in the universe is revolving around us.

Today we know that this is not true. We know what we see is a phenomenon created by the rotation of the earth itself. However, there are other phenomenon that are just as misleading and that still exist today in our vision of the subatomic particles that all matter is made out of. One of the most important of these is the electrostatic charges of the electron and the proton.

"Electrostatic" charges:

The very word "electrostatic" is misleading. It seems to suggest that the charges surrounding these particles are "static", unmoving. This is a mistake. This is a phenomenon that has created some disastrous assumptions as to how the particles of nature are created.

Contrary to popular belief, these charges are anything but static. They are dynamic. They consist of three dimensional space flowing into and out of three dimensional holes we mistakenly call the particles of nature. As seen in Books 1 & 2, it can now be understood that these "particles" are in fact holes in space. That space flows into and out of these holes creating the charges of nature.: a revolutionary vision that now allows us to leap beyond the mistaken beliefs falsely generated by 20th Century science and progress onward into a new era of understanding.

Although it is just a simple transposition of ideas to go from static to dynamic, the results are shocking to mainstream beliefs that other ideas are based upon. For example:

Lines of flux:

These lines are not static lines as is presently believed. And the belief that they are static reveals one of science's greatest mistakes when it comes to electrons.

Today's scientists have had much education. Many are mathematical marvels. They can analyze many seemingly impossible problems using mathematics. However, all of this expensive, and expansive education has not allowed them to solve one of the simplest yet most unanswerable questions children in a grammar school class might ask them: what creates the large electrostatic charge of the electron?

Although most do not know it, *no scientist in the world* [except those who will be trained in the Vortex Theory of Atomic Particles] can answer this simple question!

The scientists of today are just as ignorant of the creation of the charge on the electron as the men of old were about the shape of the earth. A quick review for those not "educated scientists" reveals that the charge on the little electron is just as large, yet of opposite value to the charge on the more massive proton.

According to classical physics, the charge on the proton is created by the addition of the partial charges of the quarks in its interior. This seems logical until we come to the electron and its anti-particle the positron. Both of these "particles" are so small, both are considered to be "point particles". [Meaning that they are so incredibly tiny, they are considered to be like pin holes without anything inside of them.] However, how come they possess the same size charge as the more massive particles like the protons that contain quarks? And how come the anti-particle of the electron, the positron, with no internal structure, has the exact same valued charge as the more massive proton with its internal structure made out of quarks that are supposedly responsible for creating its electrostatic charge? Shockingly, the greatest scientists in all of the most distinguished universities of the world, including all the Nobel Prize winning physicists of the past century cannot answer this simple question!

Coulomb's charges!

They are still useful in solving mathematical equations but now are understood to be created by flowing three dimensional space and not "static" charges at all. In the future scientists will think of them as being electro*dynamic* charges and not electro*static* charges.

Louis de Broglie: the electron is like a wave

This is very close to the truth. However, it is the denser volume of space surrounding the electron that causes surrounding space to expand and contract as the electron passes by that makes it seem like a wave. As explained in Book 1, it is the three dimensional hole in space that creates the particle effect, and the surrounding volume of denser space that creates the wave effects.

Erwin Schrödinger: quantum mechanical model of the atom

In 1926 Erwin Schrödinger, an Austrian physicist, took the Bohr model of the atom one step further. Schrödinger used mathematical equations to describe the likelihood of finding an electron in a certain position. This atomic model is known as the quantum mechanical model of the atom. Unlike the Bohr model, the quantum mechanical model does not define the exact path of an electron, but rather, predicts the odds of the location of the electron. *This model can be portrayed as a nucleus surrounded by an electron cloud.* This "cloud" is broken up into the sub-shells that are presently associated with sub-energy levels.

In a sub-shell, where the cloud is most dense, the probability of finding the electron is greatest, and conversely, the electron is less likely to be in a less dense area of the cloud. This was a brilliant achievement, however, unknown to Schrödinger, when it is discovered that vortices of flowing space connect protons to electrons in atoms and not static charges, his model possesses some unforeseen astounding consequences!

Chapter 5A "Electro<u>dynamics</u>" explains the ejection of alpha particles, gamma rays, and x-rays, and electrons.

One of the great mysteries of how atoms absorb photons then expel them was solved many years ago and explained in the first book of this series, *The Vortex Theory of Atomic Particles*. Here it was discovered that photons are absorbed by the proton in say a hydrogen atom; the photon then travels through the fourth dimensional vortex and is immediately expelled out of the electron and back into three dimensional space. However, if it possesses just the right volume of three dimensional space, it can lengthen the vortex until the electron reaches a higher energy state. Here the electron stays until the atom finally expels the added volume of space in the form of a photon possessing the exact same volume of space the original photon possessed. But there is something else that can happen too! Something strange!

In some larger more massive atoms such as uranium [U238], a powerful photon called a gamma ray or an x-ray, can be expelled directly out of the nucleus of the atom! Many years ago, this did not seem to make any sense, and lent doubt as to the correctness of this theory. Because if the model of the atom created by the Vortex Theory of Atomic Particles is right, the photon should enter the proton and exit through the electron! So how can it exit through a nucleus?

Although it was probably possible to answer this question many years ago, it was not until discovering the answer to the Constant of Fine Structure, that it was realized how a more massive atom could expel a photon through its nucleus. It was a wonderful and satisfying discovery. It ended many years of doubt about the veracity of the Vortex Theory of Atomic Particles. It also solved the final problem that persisted without an answer for over 20 years. This wonderful explanation all began with Schrödinger's equation and his placement of electrons in miniature "clouds" called "shells", and sub-shells".

Schrödinger's discovery was a great advancement to the science of physics and chemistry. But his vision of the atom was incomplete. What Schrödinger did not know about was the existence of the vortices. His electrons moved about within their shells connected to the nucleus of the atom by static charges. However, when his brilliant idea of shells *and sub-shells* are combined with the Vortex Theory of Atomic Particle's *two vortices*, suddenly, the explanation is revealed why some heavier elements such as uranium U238 emit radiation directly out of their nuclei: specifically alpha particles, gamma rays, and x-rays! It is ingenious and happens like this…

As the electron moves about in its shell, and gets closer to the nucleus and the proton to which it is attached to via its vortices, the vortices shrink in length and expand in diameter. The closer the electron approaches the nucleus, the thicker the vortices become; and the volumes of the proton and the electron have to both expand to be able to absorb and expel the extra volume. As the electron moves away from the nucleus, the vortices again lengthen out and the volumes of the proton and the electron contract; see Figure 1 below…

Part 1: The vortices expand and contract in size causing the proton and the electron to do the same.

Figure 1 [for the ease of understanding, the proton and electron are shown the same size]

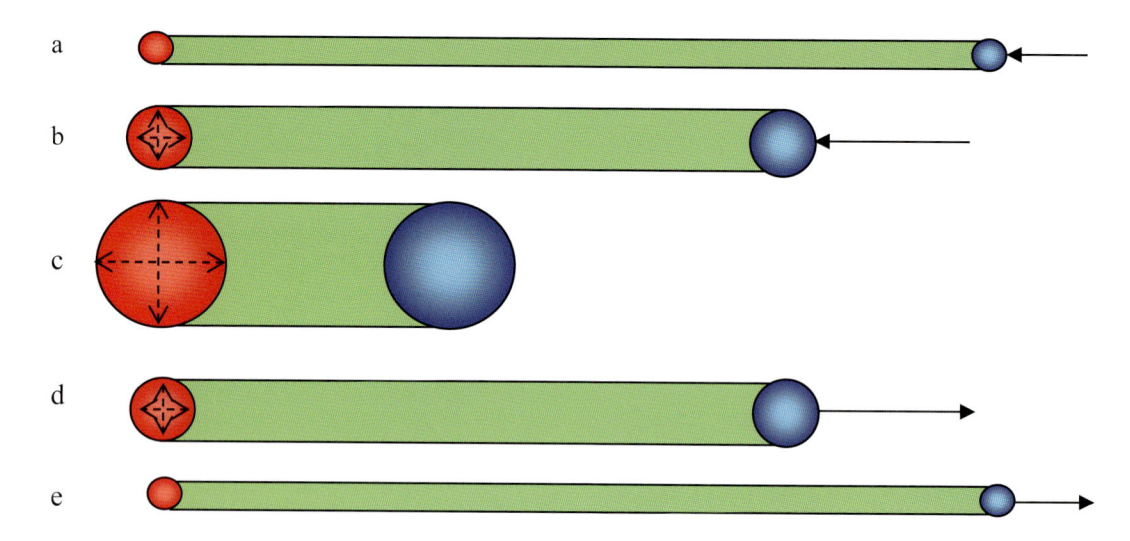

Now if the electrons in the P orbitals of the 2n shells [see Figure 2] all approach the nucleus of the atom at the exact same time, the vortices between them and the protons within the nucleus they are connected to will suddenly expand in size [see Figure 1, c above], causing the proton's volume to also expand in size…

Figure 2 P orbitals of the 2n shells

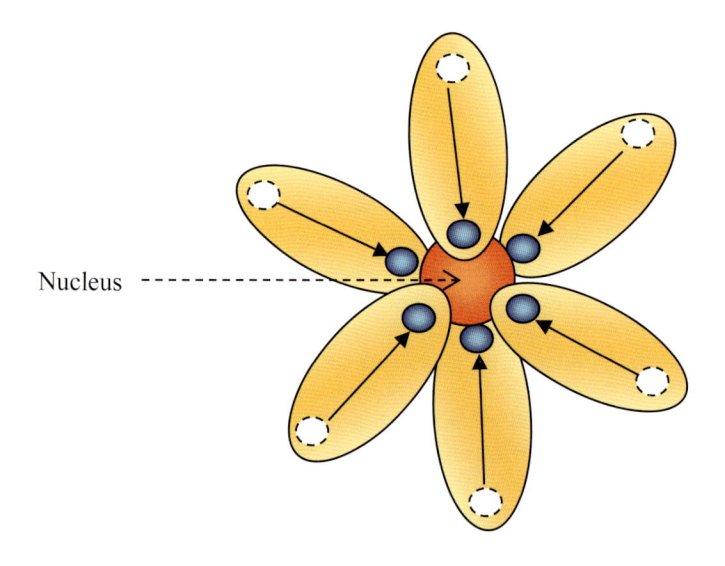

Part 2: The expansion and contraction of the proton creates a pressure wave within the nucleus.

If an alpha particle is in the center of a nucleus and is being tightly pressed on all sides by other surrounded alpha particles making it difficult to move or change positions; and if a point in time is reached where all six of the protons in the surrounding alpha particles [one for each side of the alpha particle] all expand together at the same time in phase creating a resonance; then the less dense space surrounding them and creating their nuclear gravity, generates a pressure wave [grey sphere] that pushes inward, into the alpha particle; causing this alpha particle they surround to be suddenly compressed.

Figure 3 [note: the expanding protons are actually much closer together than shown here; they are only drawn further away to allow the arrows to be inserted.] Note too: the thick black arrows represent the expanding volumes of less dense space surrounding the protons that create their nuclear gravity. This creates a gravitational wave [seen in grey] that suddenly compresses the alpha particle…

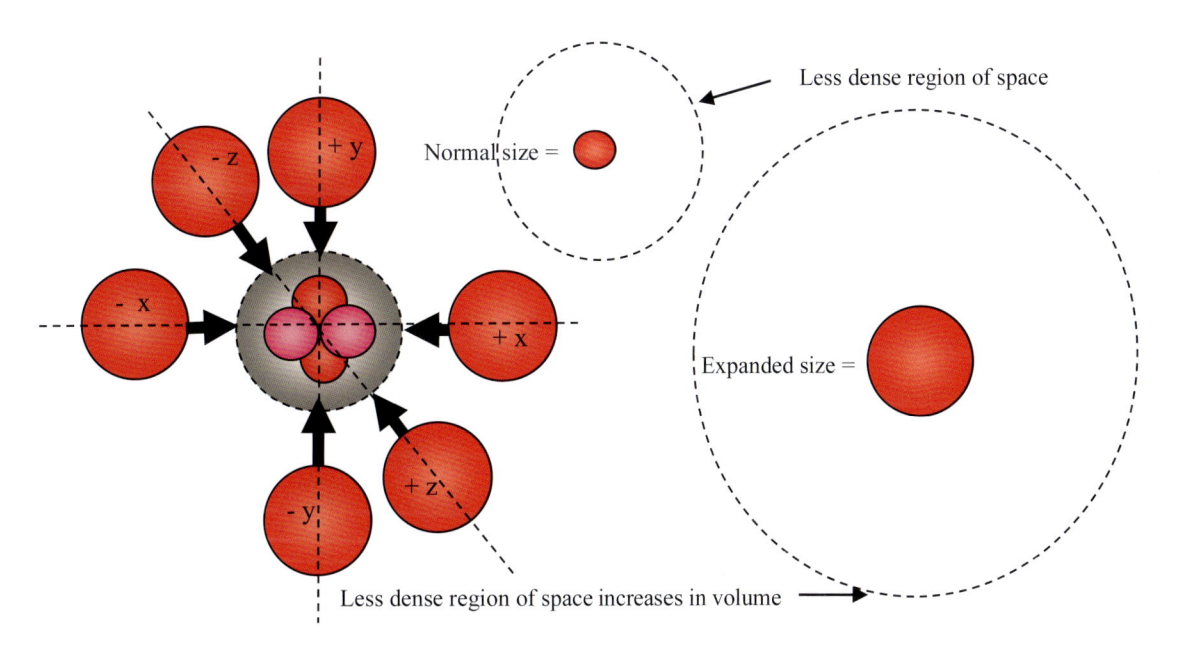

This sudden external pressure on the alpha particle causes it to be "pushed" out of the nucleus and into higher dimensional space: causing it to "tunnel" – (to move out of 3d space into 4d space); see Figure 4 below…

Part 3: The creation of the alpha particle and the gamma ray

When the alpha particle is suddenly compressed, it is thrown free of the nucleus and enters 4d space. Then, [all the following happens at the same time], since no 3d space flows into the two protons in 4d space; the two three dimensional vortices connecting the two electrons to the two protons in the alpha particle continue onwards to where the protons once were; their three dimensional volumes are shot into the void where the alpha particle was and then out through the nucleus and into the surrounding 3d space, becoming either gamma rays or x-rays. If the vortices are long, their volumes will be large and they will form gamma rays; if the vortices are short, their volumes will be smaller and they will form x-rays: the alpha particle is also shown here; it first enters 4d space at the speed of the compression wave from the six protons, passes "underneath" the nucleus, then returns back into 3d space almost instantly at the same speed; the inward flow of the surrounding 3d space into the two protons recreates its "charge"; that in turn, flows back into the electrons via the 4d vortices, reestablishing their outward flowing "charges".

[Note: the 3d vortices in red represent the expulsion of their volumes that have now become the gamma rays.]

Figure 4

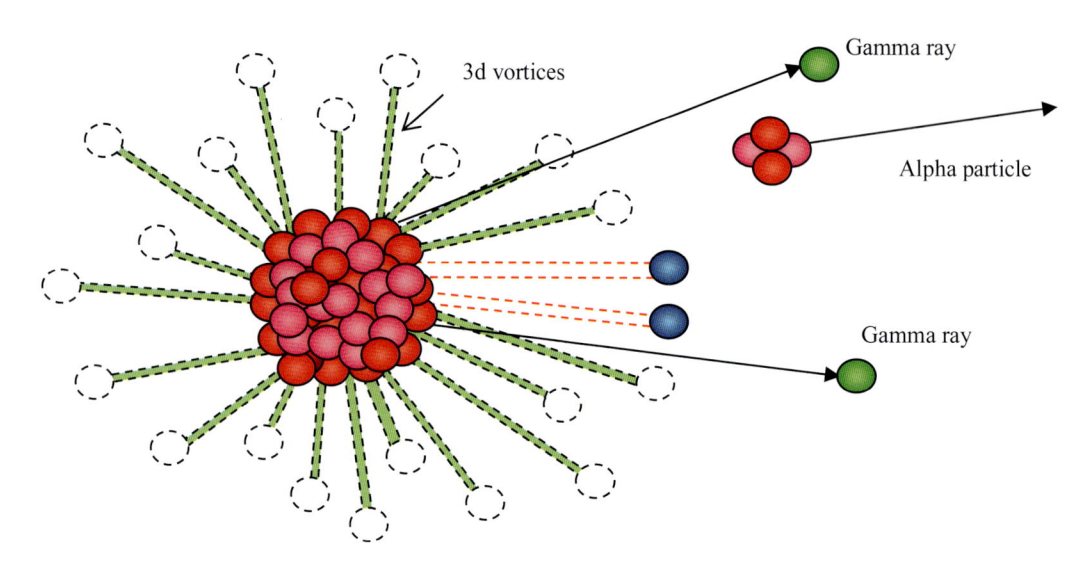

Part 4: The creation of the X-ray

If the electrons were closer to the nucleus, their vortices are shorter; hence when the Alpha particle is expelled, and the two vortices continue to flow into the nucleus, then they flow out; and being of lesser volume, become the x-rays seen in Figure 5 below…

Figure 5

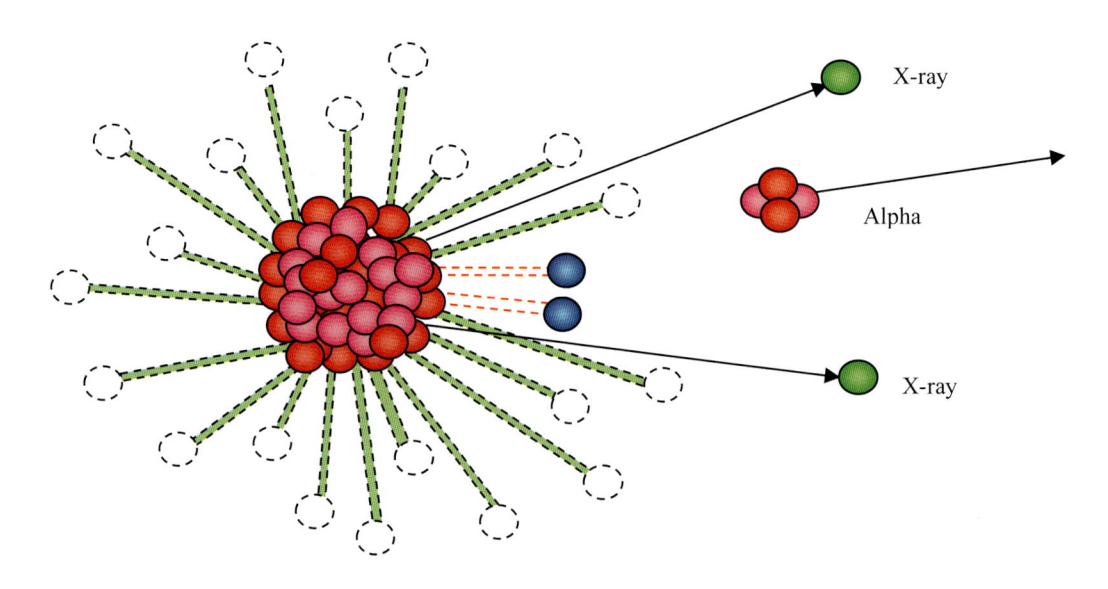

Most of what has just been said is shown in the diagram in Figure 6 below. Notice also how the two electrons in the atom and the two protons in the alpha particle are still connected via their fourth dimensional vortices [thick dotted lines].

Figure 6

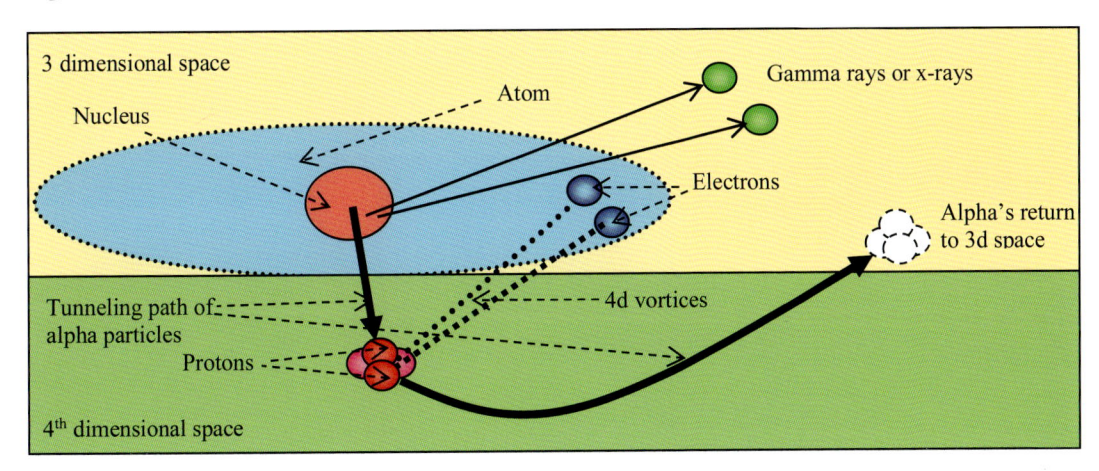

When the alpha particle returns to the three dimensional surface, its two protons are still connected to the electrons via their fourth dimensional vortices. However, they are no longer connected to the electrons via their three dimensional vortices. The volumes of three dimensional space within the three dimensional vortices that connected the electron to the proton continue on from the electron to the position in the nucleus where the two protons were; then out of the nucleus and into three dimensional space; becoming two gamma rays. However, if the electrons were in shells closer to the nucleus, their vortices would be shorter and the volumes of the photons would be less; and instead of gamma rays being emitted from the nucleus, x-rays would be emitted. Either way, the freed unattached electrons that are left, give the atom a charge of – 2, [that is if they are not also thrown free]. The 3d space flowing back into the protons gives the alpha particle its + 2 charge.

Consequently, the mechanism via which the alpha particle and gamma rays [or x-rays] are emitted from the atom can be traced to the expansion and contraction of the vortices as the electrons move closer then away from the nucleus; causing an expansion of the volumes in their interconnected protons within the nucleus; causing an alpha particle to be "compressed" out of the nucleus.

It should also be noted that this "resonance" of the six electrons happens rarely, causing the expulsion of the alpha particle to be a rare occurrence for one atom. However, because there are billions of trillions of uranium atoms in a few ounces, the U238, atoms are constantly decaying and emitting alpha particles with seemingly constant regularity.

Figure 7 [the final outcome…]

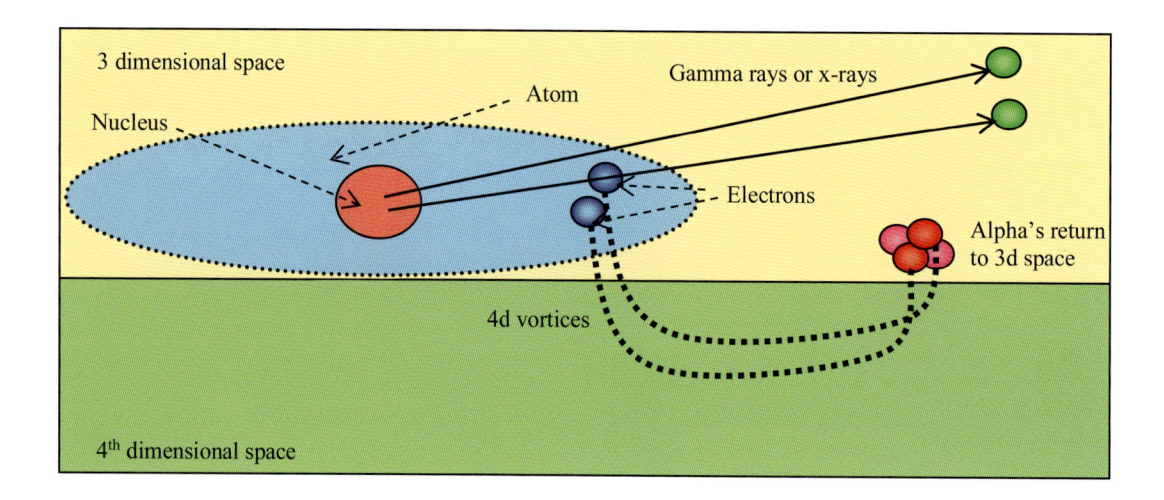

In the above Figure 7, notice that after the alpha particle returns to the three dimensional surface, its two protons are still connected to the two electrons via their fourth dimensional vortices. However, they are no longer connected in 3d space by 3d vortices; hence their "charges" return as space again begins to flow into the protons and out of the electrons. [Note: the vortices are not this curved shape but only drawn this way to show they go into and out of 4d space.]

Part 5: The expulsion from the nucleus of a "Beta" particle: now known to science as an electron.

The expulsion of the electron from a nucleus is simplistic, almost an anti-climax explanation after seeing the wonderful mechanism used for the explanation of the expulsion of the alpha particle, gamma, and x-rays. However, an explanation for all the ejections of the particles from a nucleus of an atom would not be complete without revealing how the electron is expelled too.

In the first book in this series, *The Vortex Theory of Atomic Particles*, it was explained how a proton and an electron combine to form a neutron. The neutron then lasts about 10.5 minutes in free space before decaying back into a proton and an electron. The same breakup occurs for a neutron in the nucleus of an atom only it takes much longer.

In heavier elements there are many more neutrons than protons. Consequently, the switching of identities back and forth between the protons and neutrons does not happen constantly back and forth on a one to one basis as seen in Chapter 14. Instead, this switch only occurs in long chains where as many as three protons and nine neutrons participate in switching identities over and over again. With the result being that the extra neutrons are not held as tightly together as are the paired neutrons and protons switching identities in chains of four deep within the nucleus – that are responsible for creating the alpha particles. Hence, in the course of the atom's lifetime, a time is reached where one of these neutrons is temporarily sitting alone waiting to switch into a proton when suddenly it is able to finally decay, splitting back into its two original parts: the proton and the electron.

When the neutron splits apart, the proton continues to be held in the nucleus by continuing to switch identities in the long chain of other neutrons and protons. This extra proton also changes this atom into the next highest atom in the hierarchy of the Periodic Table; while the lowly lone electron is suddenly ejected. This is how the beta particle, now known to science as an electron, is ejected from the nucleus of the atom.

References

National/International Conferences attended, and peer reviewed scientific papers presented

[1] The Vortex Theory of Matter. [Presentation of his own work]
'International Forum on New Science' Colorado State University (1992, Sept 17-20).
Moon. R. Fort Collins, Colorado. USA. Topic: The Vortex Theory of Matter. Copyright 1990)

[2] The Vortex Theory and some interactions in Nuclear Physics. [Book of abstracts; p. 259]
'The LIV International Meeting on Nuclear Spectroscopy and Nuclear Structure; Nucleus 2004' (2004, June 22-25). Moon, R., Vasiliev, V. Belgorod, Russia.
http://nuclpc1.phys.spbu.ru/nucl/Abstracts/Nucleus_2004.pdf

[3] The Possible Existence of a new particle: The Neutral Pentaquark. [Book of materials; pp. 98-104]
'Scientific Seminar of Ecology and Space' (2005, February 22). Scientific Research Centre for Ecological Safety of the Russian Academy of Sciences. Moon, R. Saint Petersburg, Russia.
https://spcras.ru/ensrcesras/

[4] Explanation of Conservation of Lepton Number. [Book of materials; p. 105]
'Scientific Seminar of Ecology and Space' (2005, February 22). Scientific Research Centre for Ecological Safety of the Russian Academy of Sciences: Moon, R., Vasiliev, V. Saint Petersburg, Russia.
https://spcras.ru/en/srcesras/

[5] Explanation of Conservation of Lepton Number. [Book of abstracts; p. 347]
'LV National Conference on Nuclear Physics' (2005, June 28-July 1). FRONTIERS IN THE PHYSICS OF NUCLEUS. Moon, R., Vasiliev, V. Russian Academy of Sciences. St. Petersburg State University. Saint Petersburg, Russia.
http://nuclpc1.phys.spbu.ru/nucl/Abstracts/Frontiers_2005.pdf

[6] The Possible Existence of a New Particle: the Tunneling Pion. [Book of abstracts; p. 348]
'LV National Conference on Nuclear Physics' (2005, June 28-July 1). FRONTIERS IN THE PHYSICS OF NUCLEUS. Moon, R., Vasiliev, V. Russian Academy of Sciences. St. Petersburg State University. Saint Petersburg, Russia.
http://nuclpc1.phys.spbu.ru/nucl/Abstracts/Frontiers_2005.pdf

[7] The Possible Existence of a New Particle in Nature: the Neutral Pentaquark. [Book of abstracts; p. 349] 'LV National Conference on Nuclear Physics' (2005, June 28-July 1). FRONTIERS IN THE PHYSICS OF NUCLEUS. Vasiliev, V. Moon, R. Russian Academy of Sciences. St. Petersburg State University. Saint Petersburg, Russia.
http://nuclpc1.phys.spbu.ru/nucl/Abstracts/Frontiers_2005.pdf

[8] The Experiment that discovered the Photon Acceleration Effect. [Book of abstracts; p. 77]
'International Symposium on Origin of Matter and the Evolution of Galaxies' (2005, Nov 8-11). Gridnev, K., Moon, R., Vasiliev, V. New Horizon of Nuclear Astrophysics and Cosmology. University of Tokyo, Japan.
https://meetings.aps.org/Meeting/SES05/Content/273
https://flux.aps.org/meetings/bapsfiles/ses05_program.pdf

[9] The Conservation of Lepton Number. [Paper presentation]
'American Physical Society 72nd Annual Meeting of the Southeastern Section of the APS' (2005, Nov 10-12). Moon, R., Calvo, F., Vasiliev, V. Gainesville, FL. USA. APS Session BC Theoretical Physics I, BC 0008
https://meetings.aps.org/Meeting/SES05/Content/273
https://flux.aps.org/meetings/bapsfiles/ses05_program.pdf

[10] The Vortex Theory and the Photon Acceleration Effect. [Paper presentation]
'American Physical Society; March Meeting; Topics in Quantum Foundations' (2006, March 13-17). Gridnev, K., Moon, R., Vasiliev, V. Baltimore, Maryland. USA.
Abstract ID: BAPS.2006.Mar.B40.6
https://meetings.aps.org/Meeting/MAR06/Session/B40.6
http://meetings.aps.org/link/BAPS.2006.MAR.B40.6

[11] The St Petersburg State University experiment that discovered the Photon Acceleration Effect.
'American Physical Society; March Meeting' GENERAL POSTER SESSION (2006, March 13-17).Gridnev, K., Moon, R., Vasiliev, V. Baltimore, Maryland. USA.
Abstract ID: BAPS.2006.MAR.Q1.146
https://meetings.aps.org/Meeting/MAR06/Session/Q1.146
http://meetings.aps.org/link/BAPS.2006.MAR.Q1.146

[12] The Neutral Pentaquark.
'American Physical Society; March Meeting' GENERAL POSTER SESSION (2006, March 13-17).Moon, R., Calvo, F., Vasiliev, V. Baltimore, Maryland. USA.
Abstract ID: BAPS.2006.MAR.Q1.147
https://meetings.aps.org/Meeting/MAR06/Session/Q1.147
http://meetings.aps.org/link/BAPS.2006.MAR.Q1.147

[13] The Neutral Pentaquark. [Paper presentation]
'International Workshop on "Nuclear Physics with RIBF' (2006, March 13-17).
Vasiliev, V., Calvo, F., Moon, R. RIKEN Research Institution. Saitama, JAPAN.
Abstract: RIBF-Pentaquark.
https://ribf.riken.jp/RIBF2006/

[14] Nuclear Structure and the Vortex Theory. [Paper presentation]
'International Workshop on "Nuclear Physics with RIBF' (2006, March 13-17).
Moon, R., Vasiliev, V. R. RIKEN Research Institution. Saitama, JAPAN.
Abstract RIBF-Vortex
https://ribf.riken.jp/RIBF2006/

[15] Experiment that Discovered the Photon Acceleration Effect. [Paper presentation]
'International Workshop on "Nuclear Physics with RIBF' (2006, March 13-17).
Moon, R., Vasiliev, V. R. RIKEN Research Institution. Saitama, JAPAN.
Abstract Moon 1
https://ribf.riken.jp/RIBF2006/

[16] To the Photon Acceleration Effect. [Paper presentation]
'APS/AAPT/SPS Joint Spring Meeting' (2006, March 21-23).
Moon, R. San Angelo, Texas. USA. Abstract ID: BAPS.2006.TSS.POS.8
https://meetings.aps.org/Meeting/TSS06/Session/POS.8
http://meetings.aps.org/link/BAPS.2006.TSS.POS.8

[17] The Saint Petersburg State University Experiment that discovered the Photon Acceleration Effect. [Paper presentation] 'American Physical Society; Astroparticle Physic II' (2006, April 22-25).
Gridnev, K., Moon, R., Vasiliev, V. Dallas, TX. USA. Abstract ID: BAPS.2006.APR.J7.6
https://meetings.aps.org/Meeting/APR06/Session/J7.6
http://meetings.aps.org/link/BAPS.2006.APR.J7.6

[18] The Photon Acceleration Effect. [Paper presentation]
'American Physical Society; Session W9 DNP: Nuclear Theory II' (2006, April 22-25).
Gridnev, K., Moon, R., Vasiliev, V. Dallas, TX. USA. Abstract ID: BAPS.2006.APR.W9.6
https://meetings.aps.org/Meeting/APR06/Session/W9.6
http://meetings.aps.org/link/BAPS.2006.APR.W9.6

[19] The Neutral Pentaquark. [Paper presentation]
'American Physical Society; Session W9 DNP: Nuclear Theory II' (2006, April 22-25). Moon, R., Calvo, F., Vasiliev, V. Dallas, Texas. USA. Abstract ID: BAPS.2006.APR.W9.9
https://meetings.aps.org/Meeting/APR06/Session/W9.9
http://meetings.aps.org/link/BAPS.2006.APR.W9.9

[20] Controversy surrounding the Experiment conducted to prove the Vortex Theory. [Paper presentation] 'American Physical Society; 8th Annual Meeting of the Northwest Section' (2006, May 18-20). Vasiliev, V., Moon, R. University of Puget Sound. Tacoma, Washington. USA. Abstract ID: BAPS.2006.NWS.C1.9
https://meetings.aps.org/Meeting/NWS06/Content/518
https://meetings.aps.org/Meeting/NWS06/Session/C1.9

[21] The Photon Acceleration Effect. [Paper presentation]
'American Physical Society; 8th Annual Meeting of the Northwest Section' (2006, May 18-20). Moon, R., Vasiliev, V. University of Puget Sound. Tacoma, Washington. USA. Abstract ID: BAPS.2006.NWS.C1.8
https://meetings.aps.org/Meeting/NWS06/Content/518
https://meetings.aps.org/Meeting/NWS06/Session/C1.8
http://meetings.aps.org/link/BAPS.2006.NWS.C1.8

[22] Experiment that Discovered the Photon Acceleration Effect. [Paper presentation]
'International Congress on Advances in Nuclear Power Plants' ICAPP '06, (2006, June 4-8). Gridnev, K., Moon, R. Reno, Nevada. USA. American Nuclear Society.
Abstract 6006. ISBN: 978-0-89448-698-2

[23] The Neutral Pentaquark. [Paper presentation]
'International Congress on Advances in Nuclear Power Plants' ICAPP '06 (2006, June 4-8). Vasiliev, V., Calvo, F., Moon, R. Reno, Nevada. USA. American Nuclear Society.
Abstract 6045. ISBN: 978-0-89448-698-2

[24] Is Hideki Yukawa's explanation of the strong force correct?
'The International Symposium on Exotic Nuclei' Book of abstracts: Joint Institute for Nuclear Research. (2006, July 17-22). Vasiliev, V., Moon, R. Khanty Mansiysk, Siberia. Russia.
http://wwwinfo.jinr.ru/exon2006/
http://jinr.ru/

[25] The Explanation of the Pauli Exclusion Principle. [Paper presentation]
'59th Annual meeting of the American Physical Society Division of Fluid Dynamics' (2006, Nov 19-21). Moon, R., Vasiliev, V. Tampa, Florida. USA.American Physical Society;
Abstract ID: BAPS.2006.DFD.P1.17
https://meetings.aps.org/Meeting/DFD06/Content/578
https://meetings.aps.org/Meeting/DFD06/Session/P1.17
http://meetings.aps.org/link/BAPS.2006.DFD.P1.17

[26] Is Hideki Yukawa's explanation of the strong force correct? [Paper presentation]
'59th Annual meeting of the American Physical Society Division of Fluid Dynamics' (2006, Nov 19-21). Moon, R., Vasiliev, V. Tampa, Florida. USA. American Physical Society;
Abstract ID: BAPS.2006.DFD.P19
https://meetings.aps.org/Meeting/DFD06/Content/578
https://meetings.aps.org/Meeting/DFD06/Session/P1.19
http://meetings.aps.org/link/BAPS.2006.DFD.P1.19

[27] The Final Proof of the Michelson Morley Experiment; The explanation of Length Shrinkage and Time Dilation. [Book of materials] 'Scientific Research Center for Ecological Safety of the Russian Academy of Sciences: Scientific Seminar of Ecology and Space'. (2007, February 8-10). Moon, R. Saint Petersburg, Russia.
https://spcras.ru/en/srcesras/

[28] The Explanation of the Photon's Electric and Magnetic fields and its Particle and Wave Characteristics. [Paper presentation] 'Annual Meeting of the Division of Nuclear Physics Volume 52, Number 10'. (2007, Oct 10-13). Moon, R., Vasiliev, V. Newport News, Virginia. USA. American Physical Society; Abstract ID: BAPS.2007.DNP.BF.15
https://meetings.aps.org/Meeting/DNP07/Session/BF.15
http://meetings.aps.org/Meeting/DNP07
http://meetings.aps.org/link/BAPS.2007.DNP.BF.15

[29] The St. Petersburg State University experiment that discovered the Photon Acceleration Effect. 'Virtual Conference on Nanoscale Science and Technology' VC-NST. (2007, Oct 21-25). Moon, R., Vasiliev, V. University of Arkansas. 222 Physics Building. Fayetteville, AR 72701 USA.
http://www.ibiblio.org/oahost/nst/index.html

[30] The Explanation of Quantum Teleportation and Entanglement Swapping. [Paper presentation] '49th Annual Meeting of the Division of Plasma Physics, Volume 52, Number 11' (2007, Nov 12–16). Moon, R., Vasiliev, V. Orlando, Florida. American Physical Society;
Abstract ID: BAPS.2007.DPP.UP8.21
https://meetings.aps.org/Meeting/DPP07/Content/901
http://meetings.aps.org/link/BAPS.2007.DPP.UP8.21
https://meetings.aps.org/Meeting/DPP07/Session/UP8.21

[31] The Explanation of the Photon's electric and magnetic fields, and its particle and wave characteristics. [Paper presentation]
'60th Annual Meeting of the Division of Fluid Dynamics'. Volume 52, Number 12. (2007, Nov 18–20). Moon, R., Vasiliev, V. Salt Lake City, Utah. American Physical Society;
Abstract ID: BAPS.2007.DFD.JU.22
http://meetings.aps.org/Meeting/DFD07
https://meetings.aps.org/Meeting/DFD07/Session/JU.22
http://meetings.aps.org/link/BAPS.2007.DFD.JU.22

[32] The Explanation of quantum entanglement and entanglement swapping. [Poster Session]
'The 10th International Symposium on the Origin of Matter and the Evolution of the Galaxies (OMEG07) (2007, Dec 4-6) Moon, R., Vasiliev, V. Hokkaido University, Sapporo, Japan. Bibcode: 2008AIPC.1016.....S, Harvard (Astrophysics Data System) ISBN 0735405379
https://ui.adsabs.harvard.edu/abs/2008AIPC.1016.....S/abstract

Books by author {A} and work presented in other published books/booklets

1. *"The Vortex Theory of Matter"* Copyright 1990
   R. Moon. {A} Costa Mesa, California

2. *"The End of The Concept of Time"* Copyright 2000.
   R. Moon. {A} Gordon's Publications of Baton Rouge. Louisiana. ISBN 096792981-4.

3. *"The Bases of the Vortex Theory of Space"* (2002).
   R. Moon. {A} Publishing house; "ZNACK" Director Dr. I. S. Slutskin. Post Office Box 648. Moscow, 101000, Russia. p. 32. (In Russian). Journal ISSN: 2362945.

4. *"The Vortex Theory…The Beginning"* (2003). Copyright 2003.
   R. Moon. {A} (Editor's note by Prof., Dr. Victor V. Vasiliev)
   Gordon's Publications of Fort Lauderdale Fla. USA.

5. *"The Bases of the Vortex Theory"* (2003).
   Book of abstracts: Russian Academy of Sciences; ISBN 5-98340-004-5; TRN: RU0403918096768 OSTI ID: 20530263 R. p. 251. R. Moon. V. Vasiliev
   http://nuclpc1.phys.spbu.ru/nucl/Abstracts/Nucleus_2003.pdf
   http://physics.doi-vt1053.com/ISBN5-98340-004-5/Nucleus_2003.pdf

6. *"The Vortex Theory and some interactions in Nuclear Physics"* (2004).
   Book of abstracts: Russian Academy of Sciences; ISBN 5-9571-0075-7 p. 259.
   R. Moon. V. Vasiliev
   http://nuclpc1.phys.spbu.ru/nucl/Abstracts/Nucleus_2004.pdf
   http://physics.doi-vt1053.com/ISBN5-9571-0075-7/Nucleus_2004.pdf

7. *The Vortex Theory Explains the Quark Theory.* (2005).
   R. Moon. {A} Gordon's Publications of Fort Lauderdale, Florida. USA. p. 205.

8. Dr. Russell Moon PhD Thesis; *"The End of "Time"* Collection of Learned Works Addendum, 2012, (pp. 473-488) VVM Publishing House: ISBN 978-5-9651-0804-6 Editor in Chief: I. S. Ivlev. Saint Petersburg State University. St Petersburg, Russia.
   http://physics.doi-vt1053.com/ISBN978-5-9651-0804-6/Dr-Russell-G-Moon-PhD-thesis-The-End-of-Time.pdf
   http://physics.doi-vt1053.com/ISBN978-5-9651-0804-6/Natural_Anthropogenic_Aerosoles_4pages.pdf

9. *"The Discovery of the Fifth Force in Nature: The Anti-Gravity Force"* Collection of Learned Works (pp. 489-495) R. Moon. M. F. Calvo.
   VVM Publishing House: ISBN 978-5-9651-0804-6 p. 534. Editor in Chief: I. S. Ivlev. St Petersburg State University. St Petersburg, Russia.
   http://physics.doi-vt1053.com/ISBN978-5-9651-0804-6/The-Discovery-of-the-Fifth-Force-in-Nature:-The-Anti-gravity-Force.pdf
   http://physics.doi-vt1053.com/ISBN978-5-9651-0804-6/Natural_Anthropogenic_Aerosoles_4pages.pdf

10. *"The Discovery of the Fifth Force in Nature: The Anti-gravity Force"* Collection of Learned Works (pages 496-503) V. Vasiliev. R. Moon. M. F. Calvo. VVM Publishing House; ISBN 978-5-9651-0804-6 2013. p 534. Editor-in-Chief: I. S. Ivlev, Saint Petersburg State University. St. Petersburg. Russia.
http://physics.doi-vt1053.com/ISBN978-5-9651-0804-6/The-Discovery-of-the-Fifth-Force-in-Nature:-The-Anti-gravity-Force.pdf

Other References

1) Stephen Smale. '*A Classification of Immersions of the Two Spheres*'. Transactions of the American Mathematical Society, Vol. 90, No. 2. (Feb 1959). Online; ISSN 1088-6850. Printed version; ISSN 0002-9947
2) Rucker, Rudy. '*The Fourth Dimension: Toward a Geometry of Higher Reality*'. (1984) ISBN-10: 0486779785
3) Abbott, Edwin, A. *FLATLAND*: '*A Romance of Many Dimensions*'. New York, Dover Publications. (1953) ISBN-13: 9798630248015
4) Besancon, Robert M. '*The Encyclopedia of Physics*'. Pages 568-570; and 949-950. Van Nostrand Reinhold Co. (1974) ISBN-10: 1124004475
5) Condon, E. U. '*The Handbook of Physics Second Edition*'. McGraw Hill Book Co. (1967) ISBN-10: 0070124035
6) Eisberg, Robert. '*Fundamentals of Modern Physics*'. Pages 9-15. John Wiley and Sons. (1961) ISBN-10: 047123463X
7) Dorothy Michelson Livingston. '*The Master of Light*'. A biography of Albert A. Michelson. University of Chicago Press. (1973) ISBN-10: 0684134438
8) Christopher Scarre. Smithsonian Institution; '*Smithsonian Timelines of the Ancient World*'. Page 65. Published September 15th 1993. ISBN-10: 1564583058
9) R. W. Heath, R. R. Macnaughton, D. G. Martindale. 'Fundamentals of Physics'. D.C. Heath Canada Ltd. (1983) ISBN-10: 0663008243
10) Cowles Encyclopedia of Science, Industry, and Technology. Cowles Book Company, Inc. (1969). ISBN-10: 0402010817
11) Christine Sutton: '*Spaceship Neutrino*'. Cambridge University press (1992) ISBN-10: 0521367034
12) Yuval Ne'eman, Yoram Kirsh: '*The Particle Hunters*'. Cambridge University Press (1996) ISBN-10: 0521476860
13) Cindy Schwarz. '*A Tour of the Subatomic Zoo*'. American Institute of Physics (1992) ISBN-10: 1563966174
14) F. Close, M. Marten, C. Sutton. '*The Particle Explosion*'. Oxford University Press (1987) ISBN-10: 0158535551
15) Robert Jastrow, Malcolm Thompson. '*Astronomy: Fundamentals and Frontiers*'. John Wiley & Sons, Inc. (1972) ISBN-10: 0471440787
16) Richard P. Feynman. '*The Strange Theory of Light and Matter*'. Princeton University Press. (1985) ISBN-10: 069116409 ISSN-13: 9780691164090
17) William J. Thompson. '*Angular Momentum: An Illustrated Guide to Rotational Symmetries for Physical Systems*' (1994) Printed version; ISBN 9780471552642 Online; ISBN-10: 9783527617821
18) Sir Fred Hoyle. On '*Nuclear reactions occurring in Very Hot Stars: The synthesis of the elements from carbon to nickel*' (1954) Astrophysics. J. Suppl. 1, 121-146
19) Alister McGrath. '*A Scientific Theology*' Volume 3; p 197. T & T Clark. New York. (2003) ISBN-10: 0567083497
20) Evgeny Epelbaum, Hermann Krebs, Timo A. L¨ahde, Dean Lee and Ulf-G, Meißner. '*Dependence of the triple-alpha process on the fundamental constants of nature*' page 15. (Sept 2013)   arXiv:1303.4856v2

Russian Scientific Journals:

1. http://www.new-philosophy.narod.ru/RGM-VVV-RU.htm (in Russian)

2. http://www.new-philosophy.narod.ru/MV-2003.htm (in English)

3. http://www.new-idea.narod.ru/ivte.htm (in English)

4. http://www.new-idea.narod.ru/ivtr.htm (in Russian)

5. http://www.new-philosophy.narod.ru/mm.htm (in Russian)

Some Subjects found in Book 3:

THE EXPLANTION OF THE QUARK THEORY

- Assessment of the Quark Theory
- Creation of the 1/3 and 2/3 Charges
- Creation of the ±1 Charge, The ±2 Charge, and Spin
- Creation of the Up and Down Quarks
- Creation of the Strange and Charm Quarks
- Creation of the Top and bottom Quarks
- "The Four Layers of Matter"
- Neutrinos Explained
- How Particle Collisions Create New Particles
- "Tunneling"
- The Explanation of how Quarks Change "Flavor"
- The Explanation of the Law of the Conservation of Lepton Number
- Lepton Creation During the Decay of Positive and Negative Pions
- Neutrino Creation During the Decay of the positive Muon
- Neutrino Creation During the Decay of the Negative Muon
- The Collision Between a Proton and an Electron Anti-neutrino & a Proton and a Muon Anti-neutrino
- The Collision between a Neutron and an Electron Neutrino & the Neutron Muon Neutrino Collision
- The Decay of the Neutron and the Creation of the Anti-neutrino
- The Explanation of the Law of the Conservation of Baryons
- The Explanation of the Conservation of "Strangeness"
- Gauge Bosons are not Force Carriers Between Particles
- The Explanation of The Pauli Exclusion Principle
- The Explanation of the CPT Theorem
- The Motion of Photons and Particles through Electric and Magnetic Fields
- The Stability of Protons; The Instability of Mesons
- Eliminating Popular Misconceptions
- Major Problems with Today's Popular Theories

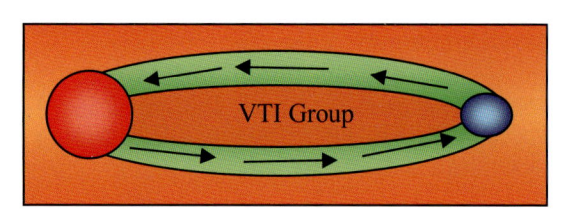

INDEX OF SCIENTIFIC DISCOVERIES FROM BOOK 3

- The Explanation of what Quarks are
- The Explanation of Quark Confinement
- The Explanation of the 1/3 & 2/3 Charges of Quarks
- The Explanation of ±2 Charge of Resonances
- The Explanation of the Up Quark
- The Explanation of the Down Quark
- The Explanation of the Strange Quark
- The Explanation of the Charm Quark
- The Explanation of the Bottom Quark
- The Explanation of the Top Quark
- The Explanation of the Muon
- The Explanation of the Tau
- The Explanation of the Electron, Muon, and Tau Neutrinos
- The Difference between Strong Force and Weak Force Creations
- The Explanation of how Quarks Change "FLAVOR"
- The Explanation of how Quarks Decay into other Types of Quarks
- The Explanation of "The Law of the Conservation of Lepton Number"
- The Explanation of the "Law" of the Conservation of Strangeness
- The Explanation of the Strange Quark's Extremely Long Lifetime
- The Explanation of the W Particle
- The Explanation of the Z Particle
- The Reason why Electric and Magnetic Fields exist in Photons and why They are at Right Angles to Each Other!
- The Explanation of the Pauli Exclusion Principle!
- The Explanation of the CPT Theorem
- The Explanation of the Asymmetric Parity of Neutrinos
- The Explanation of Gluons
- Dispelling the Myth of Gluons
- Dispelling the Myth of Gravitons
- Dispelling the Myth of the Higgs Bottom Particle
- The Explanation of the Three Color Charges in Quantum Chromodynamics
- The Explanation of the Law of The Conservation of Baryons
- The secret of quantum entanglement explained!